Construction and Design Manual
Drawing for Landscape Architects 1

About the Author
Sabrina Wilk is landscape architect and designer from Toronto. She holds international degrees in landscape architecture (Toronto), architecture (Karlsruhe) and a doctorate in art history (Dublin). Her professional design work has always focused on visual representation, both in analogue and digital graphics.

She has been a full-time professor of drawing and visualisation in landscape architecture at the Weihenstephan-Triesdorf University of Applied Sciences, Germany, since 2005, and has taught visualisation courses at many institutions across Europe. Her passion for landscape imagery led to her founding the LineScape drawing academy, and she remains dedicated to fostering a culture of diverse visualisations and graphic exploration within the landscape design professions.

Construction and Design Manual
Drawing for Landscape Architects 1

Basic Drawing, Graphics, and Projections

Sabrina Wilk

1 Drawing equipment, paper, and lines

Drawing tools
- 10 Pencils
- 11 Ink pens
- 12 Rendering
- 14 Other drawing equipment
- 15 Work space, papers, and sketchbook

The line
- 16 Pencil vs. ink
- 18 Rendering with marker
- 20 Line weights
- 24 Line types, quality, and effect
- 26 Lines and expression
- 28 Graphic possibilities

2 Projections

- 32 Introduction
- 34 Overview
- 35 Uses

3 The plan view and the rendition of symbols

Scale
- 38 Scale and drawings

Buildings
- 40 Floor plans and roof plans
- 41 Roof plans and shadows

Trees and vegetation
- 42 Drawing trees
- 48 Tree symbols
- 52 Shade, shadows, and tonal values
- 56 Tree groups
- 64 Shrubs, hedges, and grass
- 66 Trimmed hedges and woody plants
- 70 Vegetation surfaces
- 72 Flowering plants
- 76 Sketching planting beds

Built structures
- 82 Pergolas, garden pavilions, and arbours

Surfaces and materials
- 86 Paving patterns and scales
- 88 Freehand surfaces
- 90 Walls, stairs, and ramps
- 91 Rocks and stone walls

Water
- 92 Built water features
- 95 Moving water

Enliving scenes
- 96 Furniture and people

Topography and terrain
- 98 Contour lines
- 101 Retaining walls

Graphic symbols
- 102 North arrow and graphic scales

Putting everything together
- 104 Drawing process
- 106 Elements of a successful line drawing

4 Elevation and section

Elevation
- 110 Introduction
- 112 Construction

Vegetation in elevation
- 116 Trees
- 128 Shrubs and woody plants
- 129 Potted plants
- 130 Ground cover, grasses, and flowering plants
- 132 Adding depth

Built structures
- 134 Pergolas, pavilions, and arbours
- 136 Walls and materials
- 138 Water
- 139 People

Section and section-elevation
- 140 Introduction
- 142 Sections through buildings
- 144 Section cut lines
- 145 Constructing a section
- 146 Uses and scales
- 148 Section cut area
- 150 Examples

5 Parallel projections

Parallel projections
158 Introduction
159 Isometric

Elevation oblique
160 Introduction and construction

Axonometric projection
161 Introduction and construction
162 Circles
163 Trees
164 Vegetation
166 Construction steps

6 Perspective

Perspective projection
172 Introduction
174 Characteristics
178 Vanishing points
180 Types of perspective
182 Coordinates and sightlines
183 Cone of vision
184 Constructing a perspective grid
185 Diagonals
186 Horizon line and pictorial effect
188 Stairs and ramps
189 Reflections
190 Repetitive forms and dimensions

Construction methods
194 From the plan view
198 Using a perspective grid
204 From photos
206 Drawing freehand perspectives
208 Estimating proportions
210 Freehand one-point perspective
212 Freehand two-point perspectives
214 Atmospheric perspective
216 Graphic emphasis

7 Architectural presentations, layout, and lettering

Layout
220 Introduction
222 Formats
223 ISO Standards (DIN)
224 Symmetry and Asymmetry
230 Montage
233 Ordering information

Adding words to a presentation
234 Text size and hierarchy
235 Key words vs. the legend
236 Hand lettering
238 Futura alphabet

8 Appendix

242 Final thoughts
243 Acknowledgements
244 Index
245 Bibliography
246 Picture credits

Foreword

"An architect must know how to lead a pen, in order to easily show through drawings what his work shall become."

Vitruvius, Roman architect and engineer, 100 B.C.

This handbook was conceived to be both instructional and inspirational. At first glance, the topic of drawing might seem a little bit dated. But even though it would appear that all today's planning projects are produced entirely by digital means, drawing and sketching remain an important part of the design and building processes in landscape architecture and other design professions. Vitruvius' remarks, made over 2000 years ago, are still valid today and landscape architects continue to draw and put ideas down on paper to communicate their visions, both with and without the help of digital media. There are plenty of professional situations where we have to communicate with a sketch – on the construction site, at public consultations and meetings in the presence of clients or other architects. The best CAD rendering skills are less useful than the confidence to hold a pencil and produce a quick sketch of what is possible.

There is so much to draw in landscape architecture. Few other professions deal with so many tasks and ideas, all of which have to be communicated in a graphic form. This wide spectrum ranges from abstract concepts, gardens and public spaces, to large-scale vistas and regions, right down to seasonal planting schemes, functional systems, ecological processes and constructional details. All these ideas have to be put down on paper, usually in very different scales and degrees of abstraction, all dependant on time available, the viewer and the project.

This book was originally designed to help students in the beginning stages of landscape architecture studies, but is also a useful reference tool for professionals in landscape architecture and other green professions. The first seven chapters are instructional, outlining both the projections as well as the different possibilities with which to draw them. These chapters are a collection of different styles and approaches for students to look at and perhaps emulate as they learn to apply the projections. Unlike students of architecture or other design degree programmes, the majority of my students in landscape architecture start out without any skills or previous experience in drawing. There is also less and less time being allocated to teaching drawing, making it more and more difficult to acquire the drawing skills needed to communicate effectively. The final chapter is the inspirational part of the book, with an extensive collection of eye-catching drawings and visualisations from different international landscape offices. These images underscore the message that sketching and drawing still have an important role to play in day-to-day professional practise.

The book emphasises the basics of drawing, focusing on orthographic projections and line drawings, which endure as the foundation of architectural graphics, regardless of media. The black-on-white sketches are the common denominator of graphic communication, as a starting point for the description of an idea, its abstraction and representation on paper, and all of the stages of alteration and perception that come within the process of thinking and designing. A line drawing's strength lies in its directness, its ability to distil information and communicate it in an economical way.

The many different sketch examples included in the book, some of which have been digitally enhanced, are intended to inspire and encourage students to explore basic techniques and, I hope, move on to find their own graphic expression when developing their projects. It is becoming increasingly difficult to convince students, who are used to instant digital results, that they need to take the time to engage with the act of drawing, particularly since this process is an important counterpart to the instant gratification afforded by the computer. Drawing and developing one's own graphic language do require time, devotion and exploration. They can depict approaches, perception and thought processes in a way that the computer simply cannot.

Most importantly, drawing in landscape architecture is a wonderful combination of expression and communication. It is a link between hand, eye and brain, and, as such, it can be deeply personal. Even though our drawing is bound to taught representational systems and methods, everyone inevitably develops their own unique way of expressing themselves. A look back into landscape architectural graphics in the twentieth century proves this. At no other time were there so many different personalities reflected through a myriad of graphic styles, each embodying their own unique understanding of a landscape design. Although landscape architectural graphics now lie largely in the realm of digital media, hand drawing still remains a valuable part of externalising a response or approach to a design scheme.

When beginning my courses, students often lament that drawing is frustrating and does not lead to the results they want. Drawing needs time, motivation and continuity. However, those who do put perfectionist tendencies aside, stay patient and have the courage to make mistakes, quickly discover that drawing is not only effective, it is fun. The more often one draws, the more proficient one becomes and, most importantly, the more enjoyment one has doing this marvellously diverse, communicative and expressive activity. I wish my readers much enjoyment and inspiration.

Drawing equipment, paper, and lines

Drawing tools
10 **Pencils**
11 **Ink pens**
12 **Rendering**
14 **Other drawing equipment**
15 **Work space, papers, and sketchbook**

The line
16 **Pencil vs. ink**
18 **Rendering with marker**
20 **Line weights**
24 **Line types, quality, and effect**
26 **Lines and expression**
28 **Graphic possibilities**

Drawing equipment, paper, and lines

Pencils

Even in our digital age, landscape architects still sketch, design and present using pencil and ink. These are not only cheap drawing tools, but also easily transportable and highly flexible instruments for thinking, drawing and communicating. Pencils are usually the first tool adopted to put ideas on paper. Their leads come in varying degrees of hardness. Hard leads (H) are useful for precise drawings, whereas the softer leads (B) are ideal for freehand sketching. Soft pencils can easily create lines in a variety of expressive grey scales using different degrees of pressure. These are especially useful when rendering designs and illustrations. Pencils can be used on almost every surface. Their great advantage is that lines and marks can usually be erased and changes quickly made.

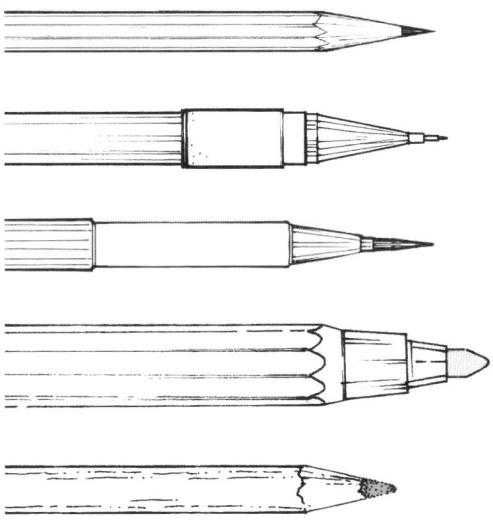
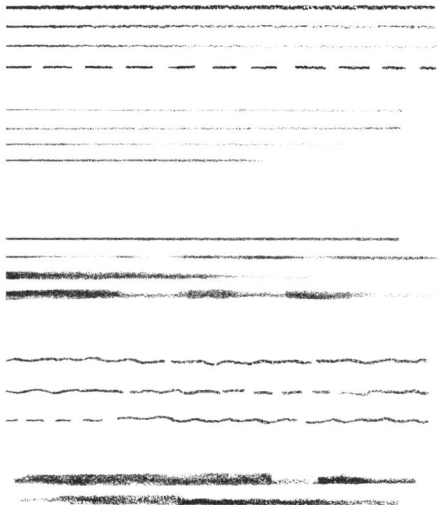

From the top:
Basic wooden pencil
Technical pencil (0,3 – 0,7 mm leads)
Mechanical pencil
Fat mechanical pencil (5 – 8 mm leads)
Eraser pencil

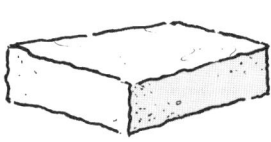
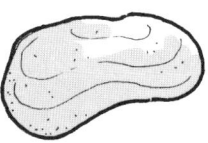

Pencil sharpenerRubber eraserKneadable eraser

Drawing tools
Pencils
Ink pens
Rendering
Other drawing equipment
Work space, papers, and sketchbook

1

Ink pens

Ink pens are also widely used in the drawing and design process. Unlike the pencil, they have defined and constant line weights, which makes them a little less flexible. Drawing with ink requires a bit more practise than with erasable pencils. Ink pens are equally useful for freehand drawing and technical drafting. They can be used on almost every surface plain paper, sketch paper, vellum, drafting films and acetates. Ink pens are usually waterproof and fade-proof, making them an ideal base for reproductions and copies. The downside to ink lines, is that they cannot be erased on paper. On vellum, they can however be removed with special erasers or by carefully scratching the surface with a razor blade.

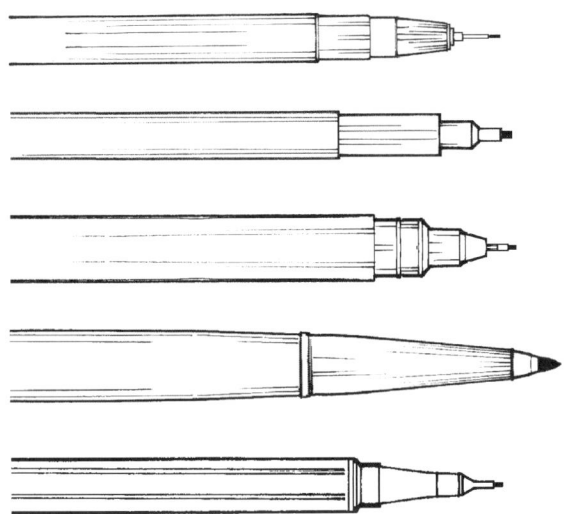
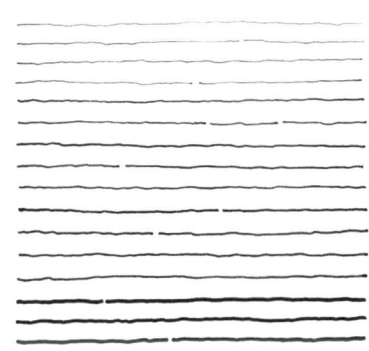

From the top:
Pigmentliner
Fineliner
Multi-liner
Felt-tip pens

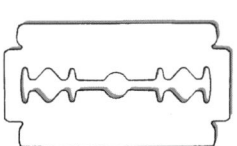

Razor blade (for use on vellum)

Drawing equipment, paper, and lines

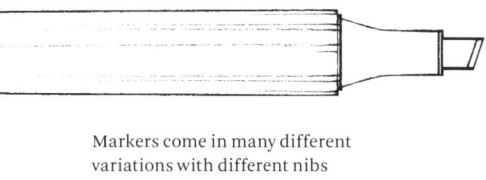

Markers come in many different variations with different nibs

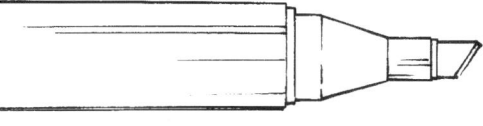

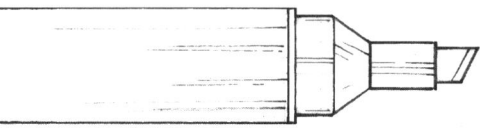

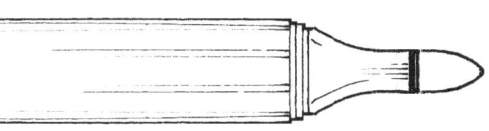

Rendering with markers
Architectural graphics have to communicate three-dimensional elements, volumes and spaces using only the two-dimensional picture plane. Black lines offer limited possibilities to create tonal effects. Softer shadows and volume effects can however be easily created with markers.

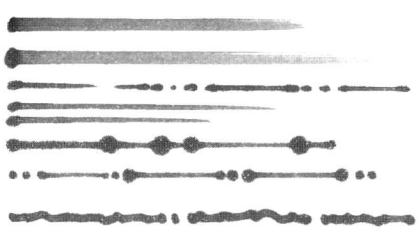

Markers come in a huge variety of colours. Their tips leave a soft layer of colour which, depending on its tonal value and saturation, is usually slightly transparent. The strokes and marks will bleed slightly on normal paper and generally get darker if they overlap each other. Markers are not meant for precision drawing. They are great tools for enhancing line drawings and adding depth effects to forms and spaces.

Drawing tools
Pencils
Ink pens
Rendering
Other drawing equipment
Work space, papers, and sketchbook

Most markers come with two different nibs. One is thin and pointed, the other has a wider and diagonal tip with a beveled edge. The thin nib allows for thinner lines, smaller marks and dots. The wide nib is excellent for larger areas of colour or grey tone, allowing for wider areas of ink pigment, sharper forms and even lettering. The two edges on the nib make it easy to quickly vary the widths of lines and strokes. Markers combine well with ink drawings, creating interesting effects to enhance simple black lines. Using markers often involves gestural movements, giving added personal effect to any drawing.

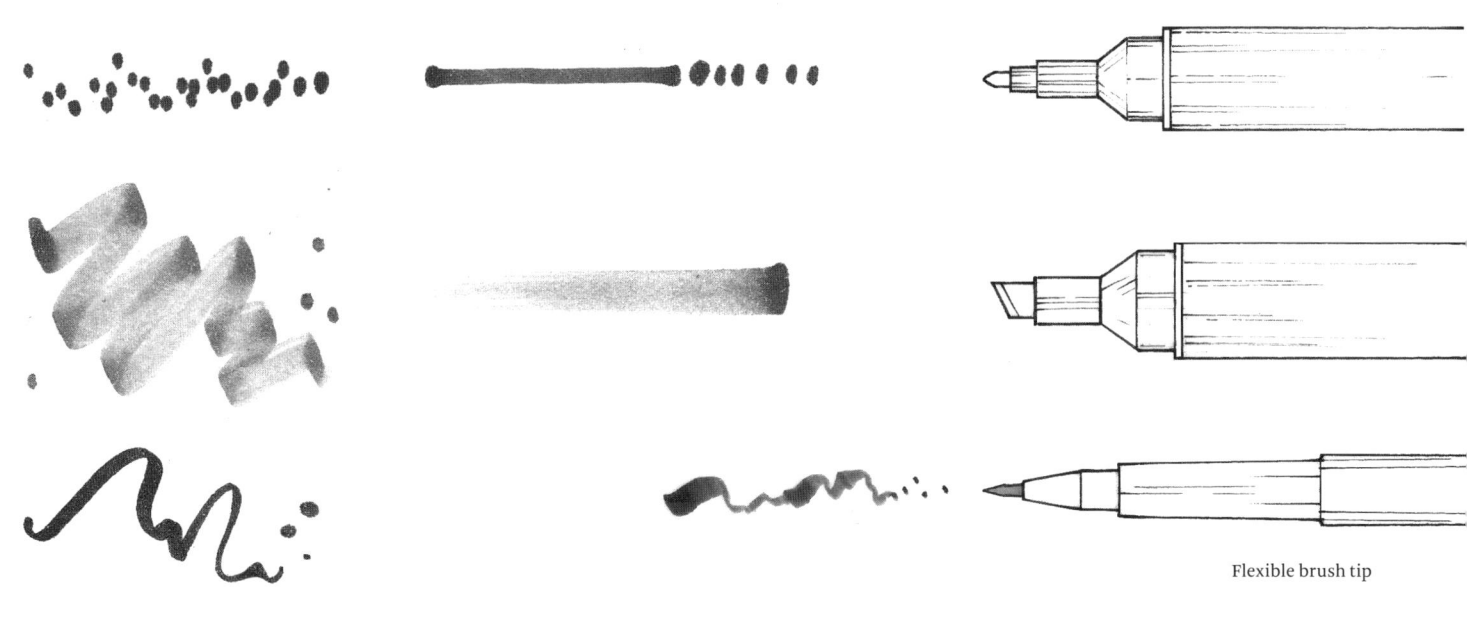

Flexible brush tip

Tapping the marker nib over the paper surface leaves a very fine diffusion of pigment, similar to an airbrush effect

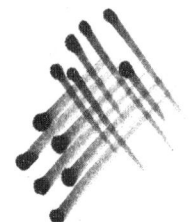

Drawing equipment, paper, and lines

Other drawing equipment

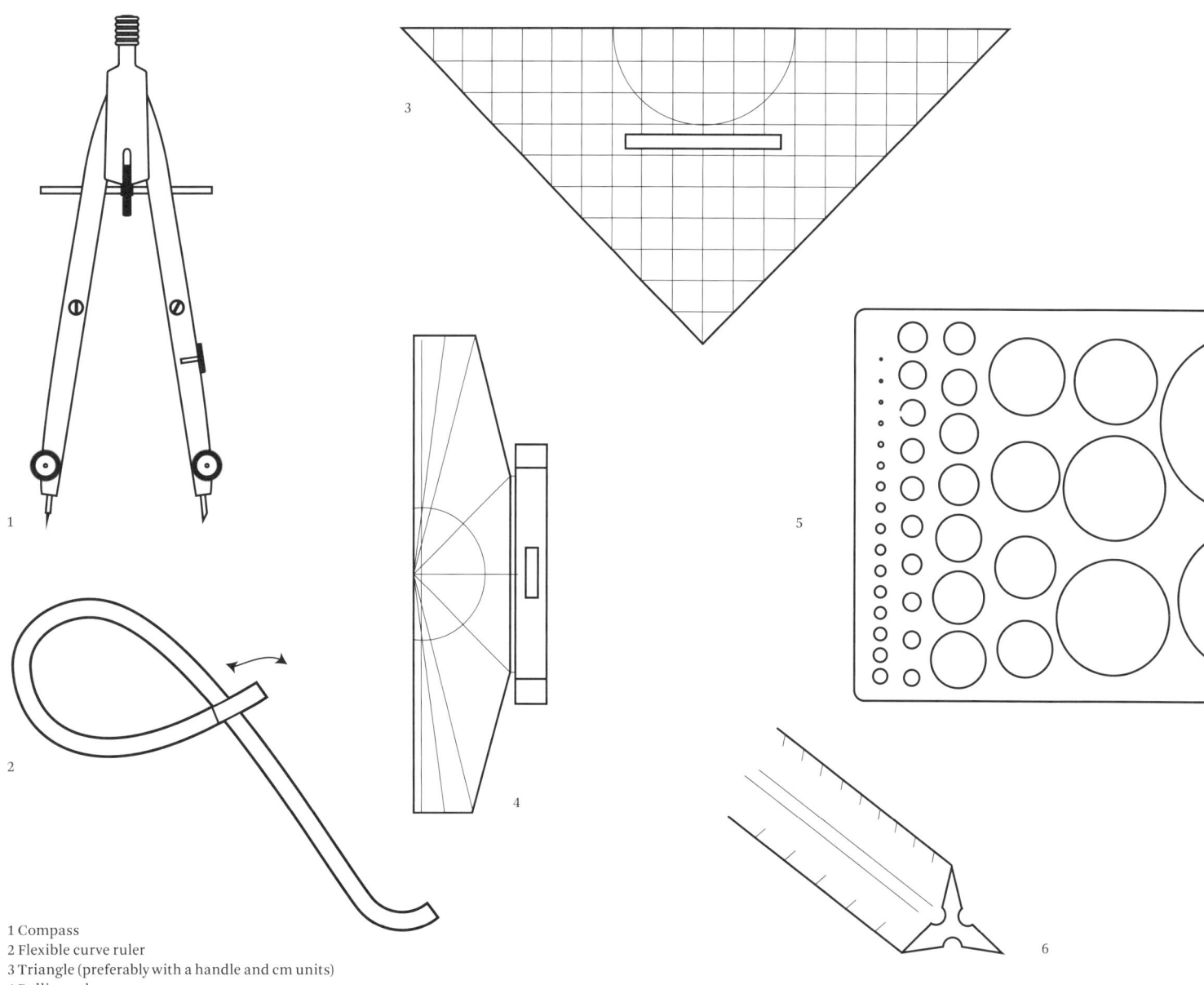

1 Compass
2 Flexible curve ruler
3 Triangle (preferably with a handle and cm units)
4 Rolling ruler
5 Circle template
6 Scale ruler (1 : 100 / 1 : 200 / 1 : 500)

Drawing tools
Pencils
Ink pens
Rendering
Other drawing equipment
Work space, papers, and sketchbook

1

Work space, papers, and sketchbook

A comfortable and well-lit work area is essential for producing good drawings and for enjoying drawing itself! A drafting table with a smooth surface and a T-square or parallel ruler is ideal. It is best to always fasten drawings well onto the table surface using masking tape, which will prevent them from sliding around and ensure precision drawing.

Sketch paper, also sometimes called tracing paper, is very thin, light (20–40 g/m²), and relatively inexpensive. It usually comes in a roll and is very useful for the initial stages of design. Sketch paper usually forms the base for thinking and designing, its transparency allowing for ideas to be layered and developed until the final design can be drafted for presentation.

Vellum is also a transparent drawing surface, however it is heavier (60–110 g/m²) and thus less susceptible to damage. Vellum is the best base for drawings that will be copied, allowing for denser and consistent line effects. It can be purchased in a roll or as individual sheets.

The sketchbook is essential for learning to draw and for gaining confidence. It is inexpensive and easily transportable and can become a very personal document to its owner. A sketchbook offers the freedom to sketch and test ideas, to observe and document spaces and details without worrying about the finished product. It is also indispensable for training perception, especially when drawing *en plein air*.

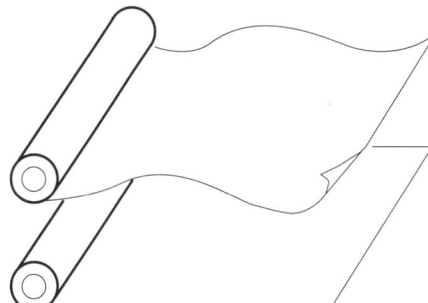

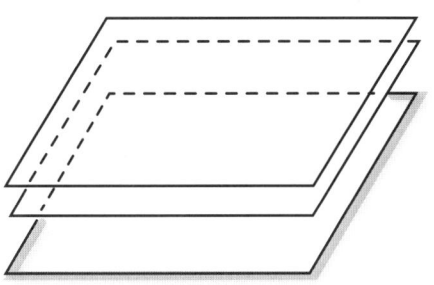

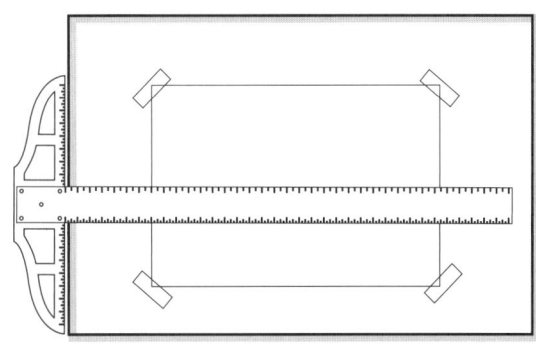

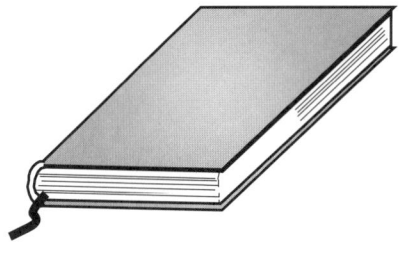

When drawing and designing, ideas are usually developed in stages and not all at once. The transparency of vellum and sketch paper allows design stages to be overlaid and layered, visibly documenting design development and its changes.

Drawing equipment, paper, and lines

Getting to know the graphic effects: pencil vs. ink
Pencil and ink pens are both equally useful when drawing, however produce very different effects. It is important to know the advantages and disadvantages of each of them in the drawing process. A simple pencil can produce a great variety of lines, strokes and tonal effects depending on different factors. The different degrees of precision and grey values are easily variable. Whether a line is precise or fuzzy will depend upon whether the pencil point is sharp or chiseled and on the degree of pressure used whilst drawing. How dark lines appear on a surface also depend upon the hardness of the pencil lead itself. This versatile drawing tool affords endless possibilities for dot lines and a large spectrum of tonal gradations of light and dark effects. It remains a firm favourite for design professionals and students alike.

Sketch elevation in pencil

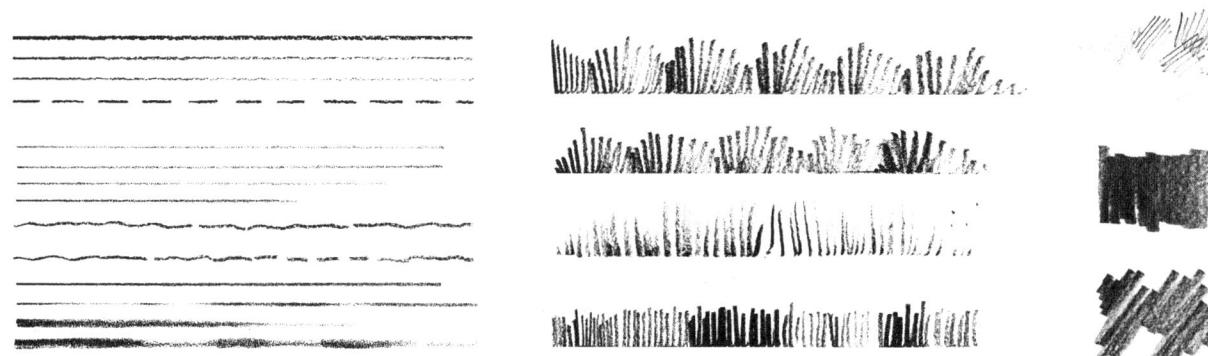

Pencil lead grades
2H – 6H (Hard), HB (Medium), 2B – 9B (Soft)

The line
Pencil vs. ink
Rendering with marker
Line weights
Line types, quality, and effect
Lines and expression
Graphic possibilities

1

It is essential to know both of these tools well to understand when to use them and what effects they can create within a drawing. Ink pens – sometimes also referred to as fine-liners or pigment liners – work only with the point tip of the pen, which determines the exact line weight produced. Unlike the pencil, adding more pressure will not change the weight or darkness of the line.

Tonal values must be produced through density of lines and strokes, as well as through hatching effects. Lines produced with ink will always have a crispness and even quality that pencil lines usually cannot achieve. They are ideal for finished work and represent an excellent basis for copying, as their thickness and blackness remains consistent and clearly legible.

Sketch elevation in ink

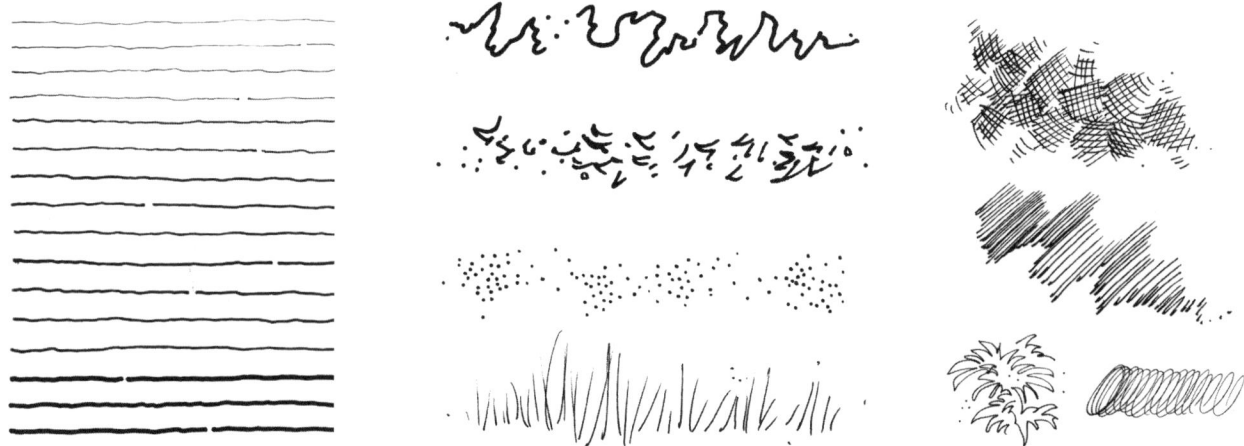

Drawing equipment, paper, and lines

Graphic effects with markers

The many variations of strokes, textures and hatchings made possible by markers are great for enhancing a black-line drawing. It is important to remember that their tonal values will appear darker on paper than on vellum, and that hand movements and gestures must be considered as part of the overall rendering effect.

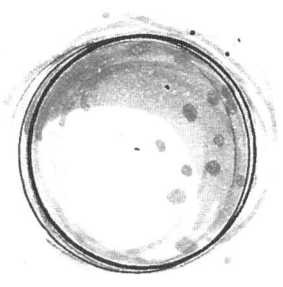

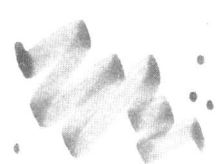

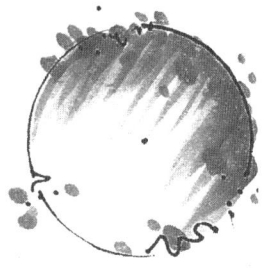

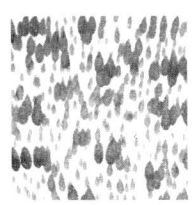

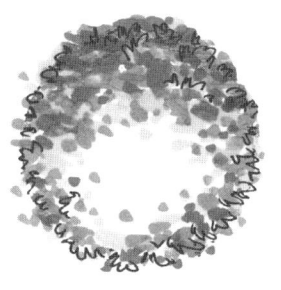

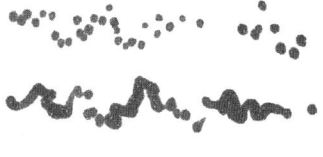

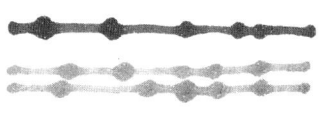

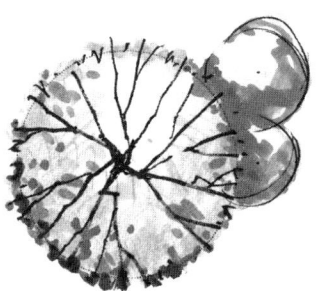

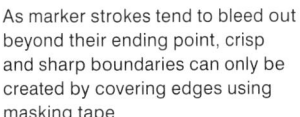

As marker strokes tend to bleed out beyond their ending point, crisp and sharp boundaries can only be created by covering edges using masking tape

The line
Pencil vs. ink
Rendering with marker
Line weights
Line types, quality, and effect
Lines and expression
Graphic possibilities

1

Each quick sketch elevation was drawn using an ink pen, then enhanced with grey marker strokes in different variations

Drawing equipment, paper, and lines

Line weights in plan and elevation

In architectural graphics, black lines usually form the basis of every drawing and projection, regardless of media. Whether drawn by hand or with the help of a digital media and CAD-programmes, lines describe forms and define planes and surfaces. In order to ensure easy legibility, they need to be brought together properly in a drawing.

The thicker the line weight in a drawing, the stronger and bolder it will appear to the viewer. Line weights need to be organised and given a hierarchy. These will differ, depending on the scale of the drawing. It is important to remember that the elements closer to the viewer can generally be drawn with a stronger line weight than those appearing further away.

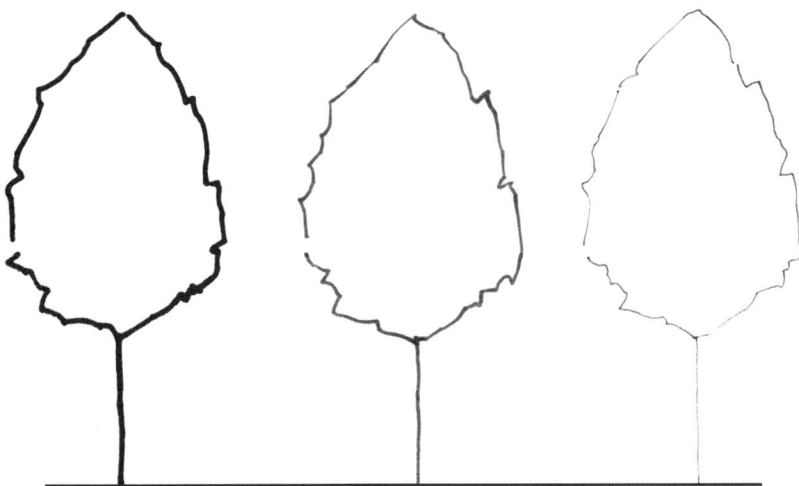

The farther away an element or plane is from the viewer, the finer the line weight it can have in a drawing. In a plan, this might be the ground plane with a paving pattern, in an elevation, this is usually the background.

The line
Pencil vs. ink
Rendering with marker
Line weights
Line types, quality, and effect
Lines and expression
Graphic possibilities

1

Note that the thicknesses shown here in this diagram on the right are indicative only. The actual line weight must be considered for every drawing in accordance with its elements and its scale.

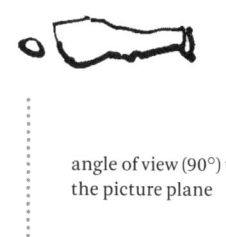

angle of view (90°) to the picture plane

Picture plane / drawing surface

Plan view

Line weights in a plan

Elevation

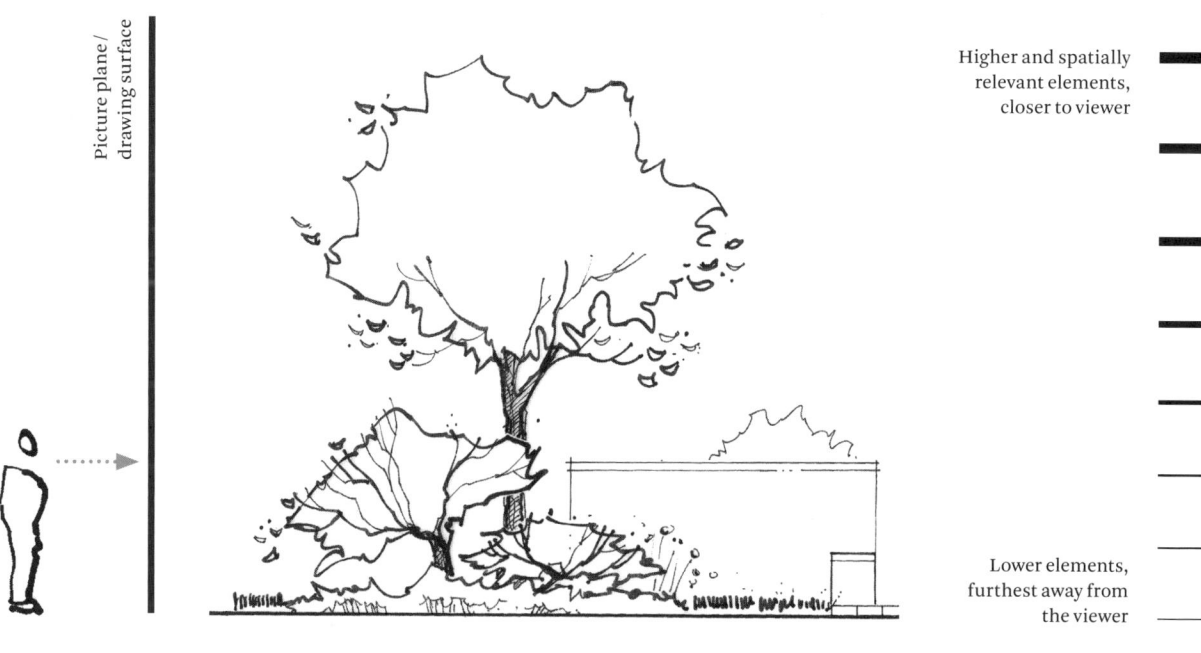

Higher and spatially relevant elements, closer to viewer

Lower elements, furthest away from the viewer

Line weights in an elevation

Foreground — Background

Drawing equipment, paper, and lines

Line weights and their effect in a drawing

Without correct differentiation of line weights in a drawing, its emphasis may become unclear and drawings difficult to read for the viewer. Different line weights help to differentiate distances from the viewer, since the heaviest lines appear closest. If the ground plane is drawn with heavier lines than the spatially defining trees or buildings, it may dominate the drawing and pull the viewer's eye towards a less important aspect of the design. Similarly, in an elevation, the foreground should carry the graphic weight in the scene as opposed to the background (*see also page 132*). When lines are close together, they often result in a textural pattern which further communicates surface materials within a plan. The resulting grey tones and patterns assist the outlines in communicating the elements to the viewer. The line weights of such textures and hatchings tend to be thinner than the form outlines. The goal of every line drawing in landscape architecture should be easy legibility, allowing the viewer to instantly understand what is happening in the space.

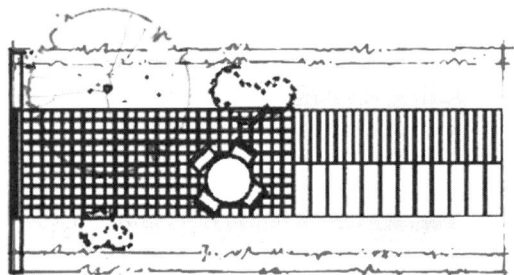

Wrong – The emphasis is on the ground plane instead than on the larger, spatially defining elements

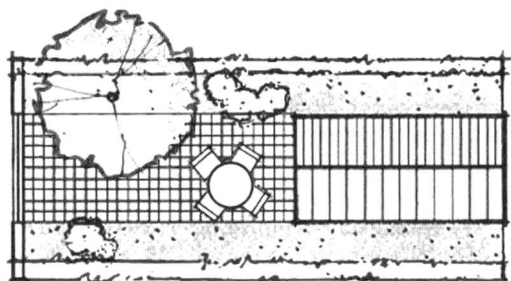

Correct – The building, tree and hedges have a stronger line weight and are read as volumes

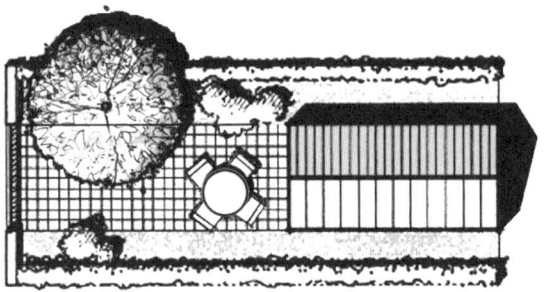

Shadows further emphasise the volumes. The contrast seemingly lifts elements from the picture plane, making them easy to read.

The line
Pencil vs. ink
Rendering with marker
Line weights
Line types, quality, and effect
Lines and expression
Graphic possibilities

1

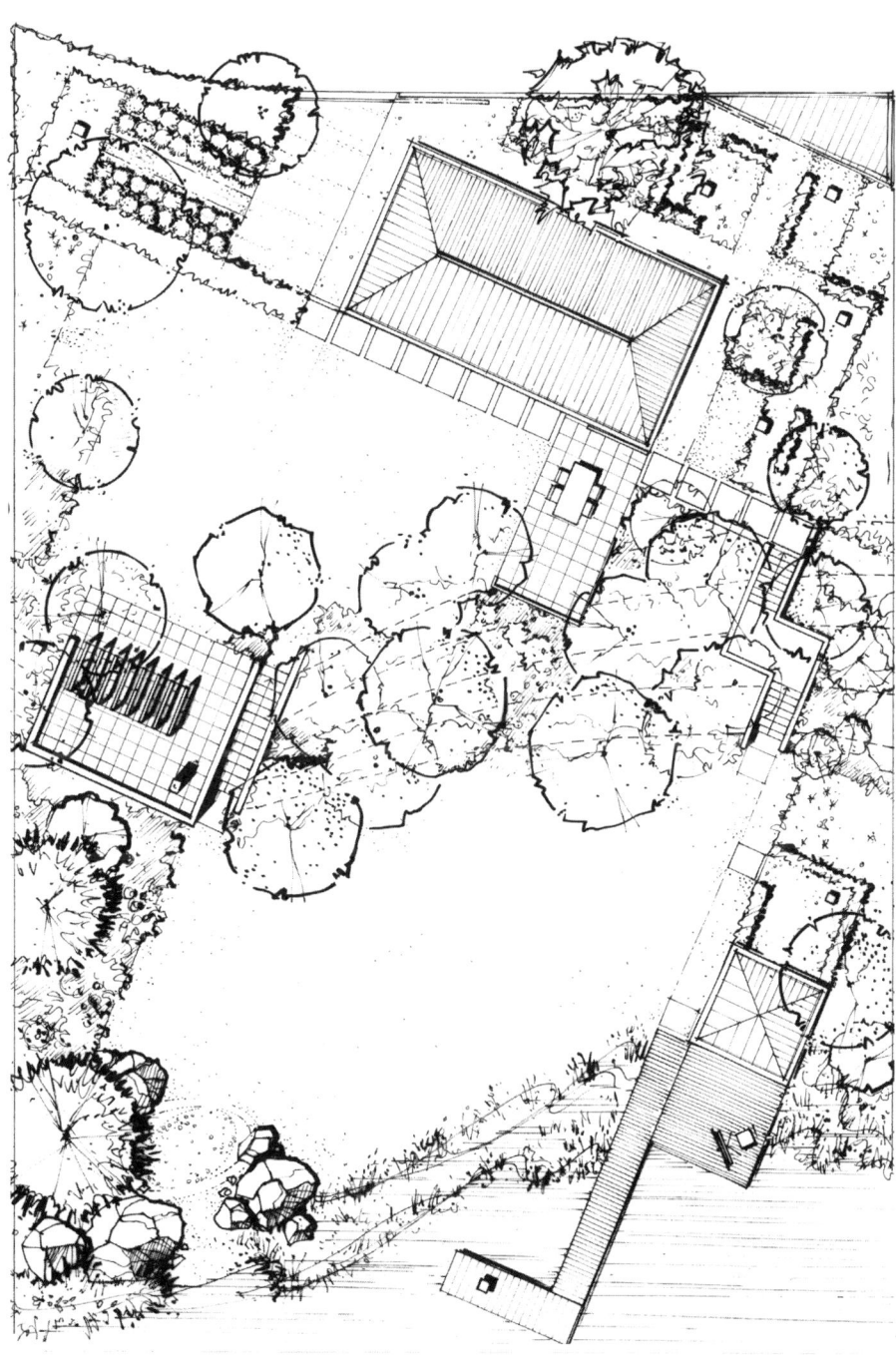

The correct assemblage of line weights allows for easy legibility of the built and vegetation components in the overall composition. Even a relatively simple line drawing, such as this plan, can communicate all of its components easily, without relying on more elaborate graphic rendering.

Drawing equipment, paper, and lines

Line types, quality, and effect
Freehand lines are usually adopted at the start of the design process. These lines are drawn without a straightedge, as imperfect but continuous strokes with a strong start and finish. Individual lines may appear a bit shaky, however, when grouped, they appear unified. These soft lines are often repeated and overlaid, especially when thinking on paper and testing ideas. Freehand lines are loose, exploratory and unfinished, often communicating work in progress.

As the design process matures, forms will become finalised, with straight lines drawn using a straightedge. These will have a slightly more finished quality. Depending on the speed with which it is drawn, the line thickness may be inconsistent and, as a result, can still appear sketchy or unfinished. This effect might be useful when presenting design variations. Overlapping corners also contributes to the sketch-like effect. The more consistent the quality of the individual lines, the more finished their appearance.

When a design is finished, more formal draftsmanship is usually required. The lines are all crisp and continuous, giving a precise and finished impression. Corners are sharp and all lines are drawn have a consistent weight throughout. These finished technical drawings are less commonly drawn by hand, as they are often left to CAD programmes, which produce perfect lines and exact corners.

The line
Pencil vs. ink
Rendering with marker
Line weights
Line types, quality and effect
Lines and expression
Graphic possibilities

1

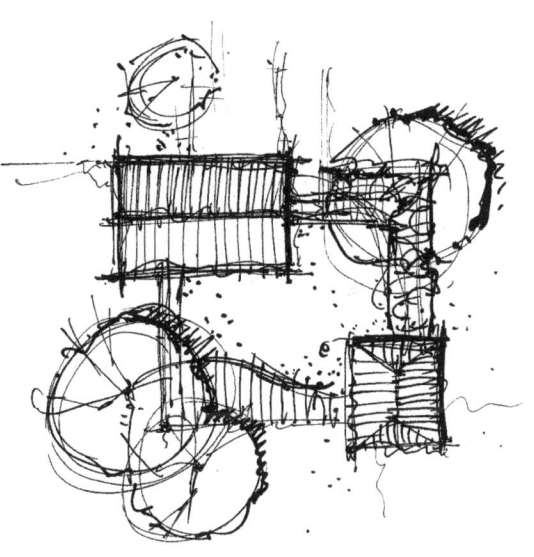

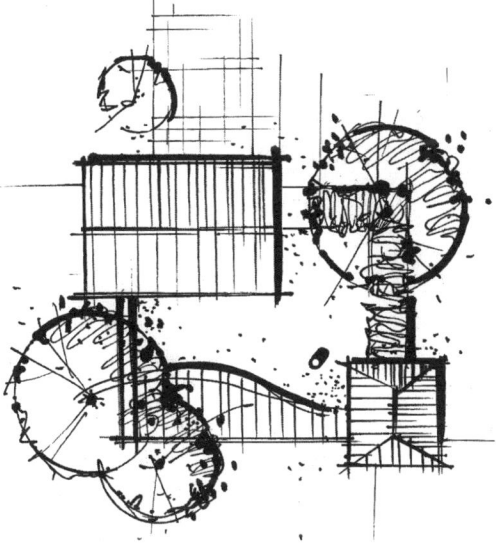

A good quality line drawing – even a freehand sketch – can be achieved through the unity and consistency of the line types used. Even if the individual lines may not be perfect, their uniformity will result in a collective graphic synthesis within the overall drawing.

Freehand lines are loose, repetitive and very useful for the initial design stage. Their imperfection affords some freedom to think and test ideas on paper, without worrying about the finished product.

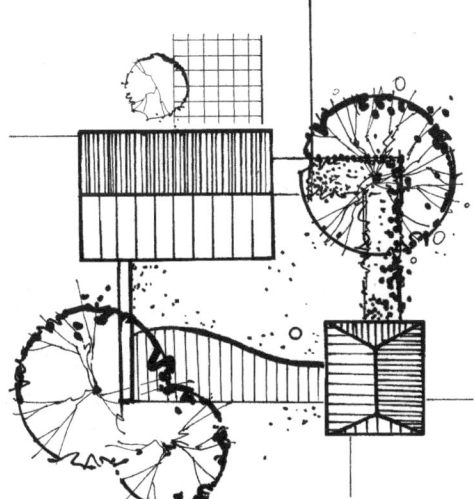

Straighter lines appear as the design progresses. Overlapping corners and variations in line quality still retain a loose, sketch-like quality, communicating a work in progress.

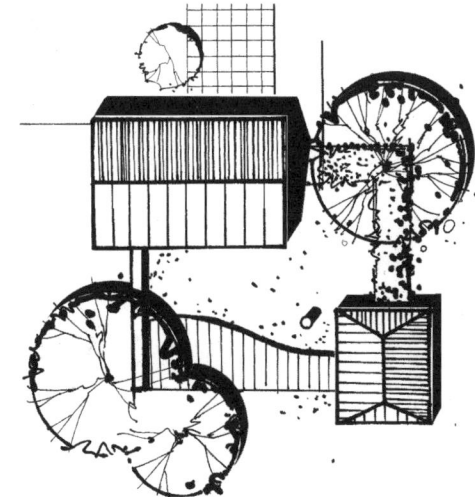

Straight, continuous and neat lines are required when every aspect of a design has been thought through and finalised. The resulting effect is finished and precise.

Drawing equipment, paper, and lines

Lines, gestures, and expression

Although landscape architectural drawings are strongly representative and have a communicative purpose, it is important to remember that drawing with lines offers great potential for creative expression. The effect of line within a drawing depends not only on its ability to define a form, but also on the distinct quality it is given by the hand of the illustrator. Drawn lines can have perceivable characteristics. Combined with movement and gestures, they can communicate more than just outlines, hatchings and volumes. Lines can communicate underlying messages. A line can appear light or heavy, calm or dynamic depending on how it is brought to paper. One only has to look at the art world to quickly understand that line drawings can suggest emotions and can bring a perceivable, very personal atmosphere to a drawing. Landscape architects are certainly not artists and their drawings underlie a strong methodology. Unlike art, the drawings are not intended for their own sake, but are almost always intended for someone else's eyes. Their goal is to reach the viewer and communicate a proposal for a design that can be translated into a built reality. However, the fact that landscape architects work with images automatically implies that they have expressive options. Each drawing reflects the hand which produced it and externalises the personal response of the individual landscape architect to the given design task.

Light

Heavy

Calm

Rhythmic

Flowing

Fast

Irregular

Shy

Lively

Hesitant

Aggressive

Slow, leisurely

Ambitious

Optimistic

The line
Pencil vs. ink
Rendering with marker
Line weights
Line types, quality, and effect
Lines and expression
Graphic possibilities

1

Lines, of course, do not exist in reality. They are a method of representation, used to distinguish and identify shapes on paper. They also can construct different textures and values in order to visually describe the seen objects. Throughout the brief history of landscape architectural graphics, it is easy to see how landscape architects have developed their own unique drawing styles and graphic languages. These were as much a product of their training, but also often an expression of their individual approach or response to a project, and a a reflection of the Zeitgeist of their day.

heavy/calm/simple

light/irregular/lively

A tree outline can be drawn in a myriad of ways, each symbol communicating a different message or perceived way of seeing the tree

unique/dynamic

Drawing equipment, paper, and lines

Drawing and sketching: Graphic possibilities
The quality of a sketch drawing is defined by the consistency of the line types it contains. Along with the tools and media used, the uniformity of the line types greatly contributes to the overall unity of a drawing, even if individual lines are not perfect in themselves. Good drawings are the sum of their graphic parts. Even the simplest sketch can be drawn in a variety of ways, each with a different effect.

How much effort goes into a drawing will depend on several factors, such as the project stage, the viewer or even the time available for production. These examples show that it doesn't take much to communicate ideas in sketch form. The more unfinished the freehand lines appear, the more room they leave for interpretation.

The line
Pencil vs. ink
Rendering with marker
Line weights
Line types, quality, and effect
Lines and expression
Graphic possibilities

1

Sketches are expressive and inexpensive, and can serve as a basis for further digital graphics. Digital graphic programmes can further enhance a hand drawing with effects which cannot be achieved using manual techniques.

The combination of digital rendering and hand drawing can produce unique effects and it is a good idea to try out as many variations as possible to achieve the desired effect for a particular drawing or sketch.

2

Projections

32 Introduction
34 Overview
35 Uses

Projections

Orthographic projections

Understanding the different orthographic projections is an essential first step in landscape architectural graphics. These are the primary graphic conventions used in architectural projects. Each projection communicates a single-view of a design and plays an important role in the development of a design and its elements. The projections are a system of representation which have been around since antiquity. They form the basis for architectural graphics and play an important role in the design and construction process. These drawings are referred to as projections, since their contents are projected orthogonally onto a picture plane.

This picture plane (the drawing surface) lies between the object and the viewer, with the images of the object projected forward to the picture planes. The viewer's line of sight also looks at the image with at the same 90 degree angle. The picture plane is parallel to the major surfaces of a building. Projecting all sides of a form or shape onto the picture plane has the advantage to avoiding any distortion of foreshortening. All shapes retain their true proportions and can be drawn to scale. This allows them to be measured, accurately which makes them extremely useful in the design and building process.

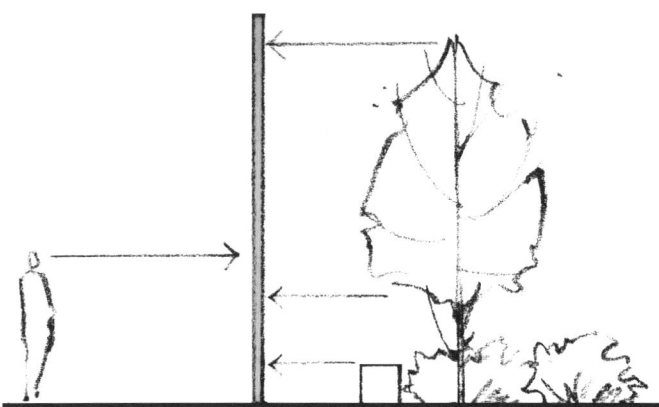

It might take some time and effort to understand how orthographic projections work and to learn to draw them. They allow us to show things to scale but not how they would normally present themselves to us. The flat projections onto the picture plane (foreground and background) give these types of drawings a high degree of abstraction.

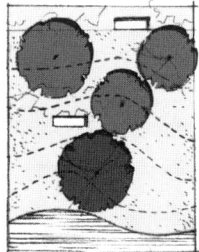  

Plan　　　　　　Elevation　　　　　　Section

Projections
Introduction
Overview
Uses

2

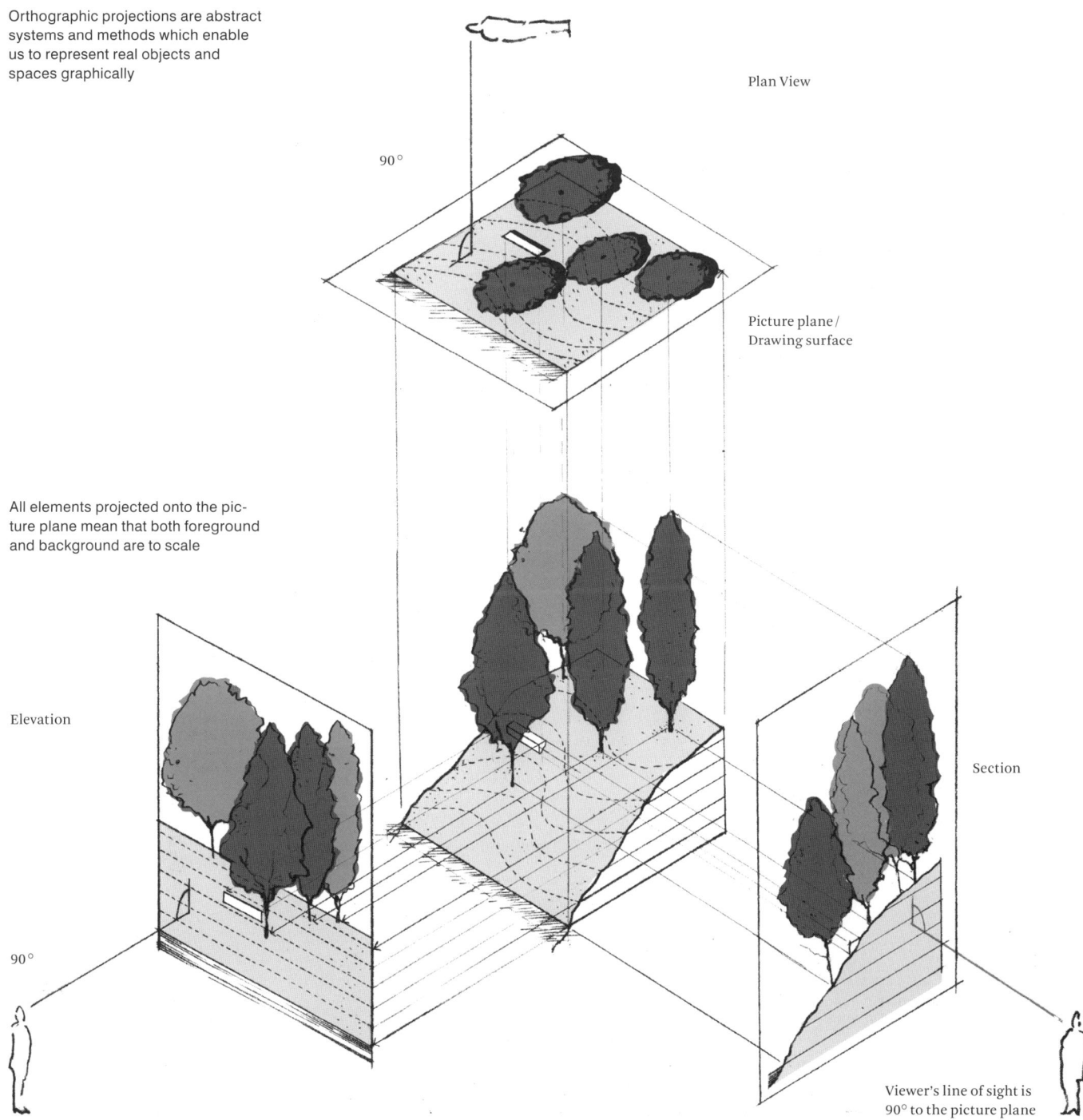

Orthographic projections are abstract systems and methods which enable us to represent real objects and spaces graphically

Plan View

Picture plane / Drawing surface

All elements projected onto the picture plane mean that both foreground and background are to scale

Elevation

Section

Viewer's line of sight is 90° to the picture plane

Projections

Overview of projection systems

Orthographic and parallel projections

Plan, section and elevation:
The primary architectural graphic conventions, showing a single-view of planes parallel to the picture plane

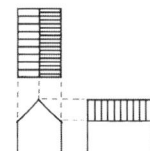

Isometric:
Three-dimensional parallel projection using three major axes, at equal angles (30° projection)

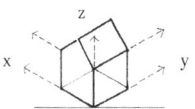

Axonometric or Plan oblique:
Parallel projection from the plan view

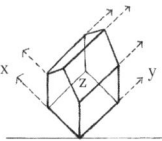

Elevation oblique:
Parallel projection from a frontal elevation

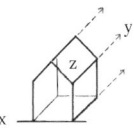

To scale and measurable / High degree of abstraction

Technical

Perspective projections

One-point perspective:
One side of an object is parallel to the picture plane; the other is perpendicular, which means its parallel lines converge in a central vanishing point

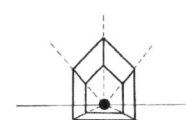

Two-point perspective:
The object is turned in relation to the viewer, with only the vertical axes remaining parallel to the picture plane. Parallel lines converge in two vanishing points.

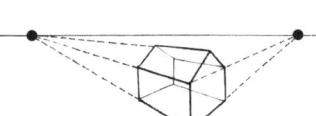

Three-point perspective:
The object is positioned so that no sides or verticals are parallel to the picture plane

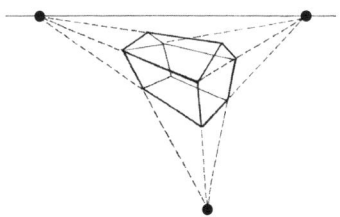

Illusional

Not to scale / High degree of presentability and very close to visual perception

Projections
Introduction
Overview
Uses

2

Uses

Orthographic and parallel projections offer a wide spectrum of representation possibilities for a design project. Landscape architects not only need to understand how projections are constructed, they also need to understand how the projection describes and communicates different aspects of an object or a space. Each method has its own graphic advantages and potential effects. Each projection also conveys a unique relationship between an object, or space, and the viewer. It is absolutely essential to know all the projections in order to effectively communicate all elements of a design project.

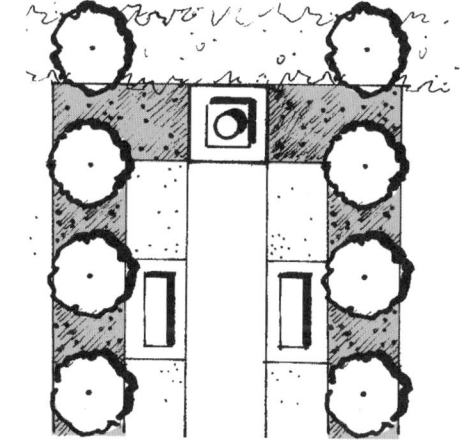

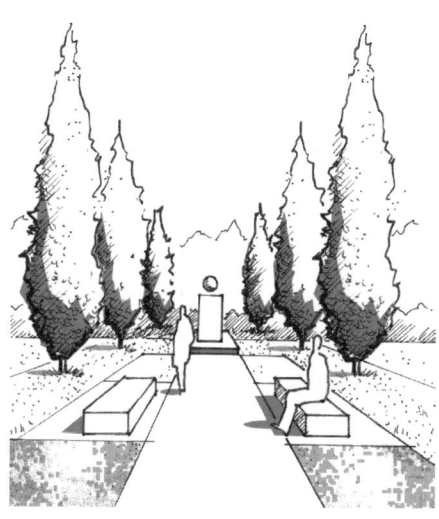

The plan view and the rendition of symbols

Scale
38 **Scale and drawings**
Buildings
40 **Floor plans and roof plans**
41 **Roof plans and shadows**
Trees and vegetation
42 **Drawing trees**
48 **Tree symbols**
52 **Shade, shadows, and tonal values**
56 **Tree groups**
64 **Shrubs, hedges, and grass**
66 **Trimmed hedges and woody plants**
70 **Vegetation surfaces**
72 **Flowering plants**
76 **Sketching planting beds**
Built structures
82 **Pergolas, garden pavilions, and arbours**
Surfaces and materials
86 **Paving patterns and scales**
88 **Freehand surfaces**
90 **Walls, stairs and ramps**
91 **Rocks and stone walls**
Water
92 **Built water features**
95 **Moving water**
Enliving scenes
96 **Furniture and people**
Topography and terrain
98 **Contour lines**
101 **Retaining walls**
Graphic symbols
102 **North arrow and graphic scales**
Putting everything together
104 **Drawing process**
106 **Elements of a successful line drawing**

The plan view and the rendition of symbols

Scales

A range of different scales is usually found when drawing and communicating a design project. The term scale refers to the proportional relationship between a real spatial dimension and its representation in a drawing. A scale of 1:100 tells the viewer that 1 cm length found in a plan or elevation represents 100 times that amount in reality, or 1 m. Legibility of forms in a plan is very important. Depending on the scale, drawn elements are often subject to a graphic simplification, reducing some elements to highly abstracted symbols.

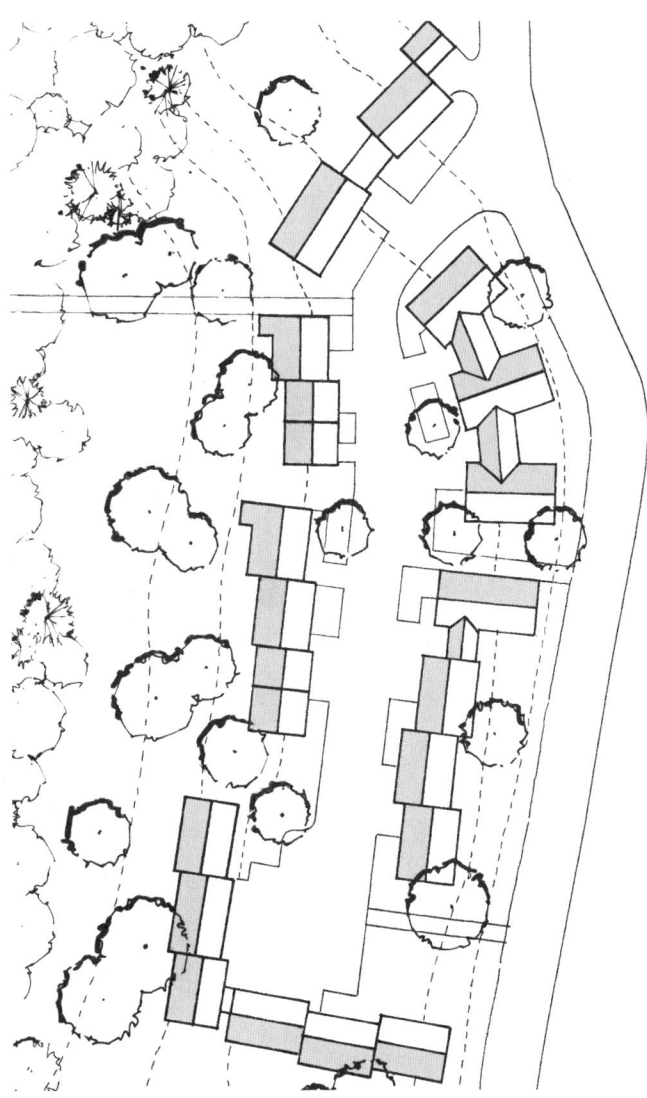

1:1,000 > 1 cm = 10 m
Plan view

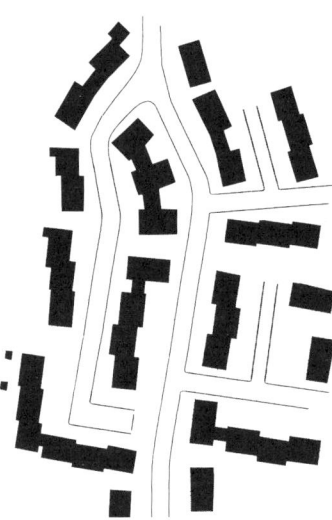

1:1,000 / 1:2,000 / 1:5,000 / 1:10,000 ...
Plan view / Figure ground plan, often found in urban and regional planning.

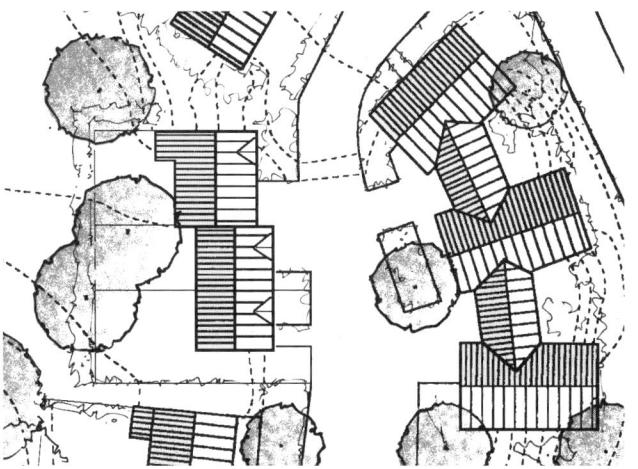

1:500 > 1 cm = 5 m

Scale
Scales and drawings

3

As the viewer zooms into a project, the scale of a drawing increases and with it the amount of detail shown. Each jump in scale should reveal different information about the project. In the design process, scales are varied to reveal different amounts of information regarding a project's contents, ranging from overall structure and order, to detailed surfaces and vegetation. Scales such as 1:20, 1:10 tend to be working and construction drawings. As these drawings are intended to give increased detail about the way a project is built, they tend to be less graphically interesting. Instead they often rely on standardised graphic norms to convey materiality and important constructive information.

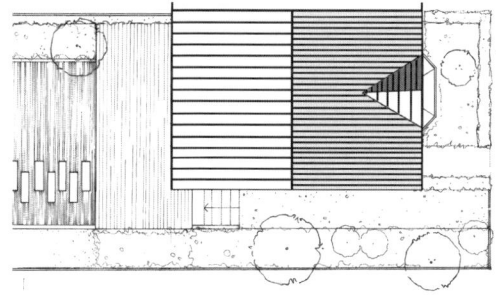

1:200 > 1 cm = 2 m

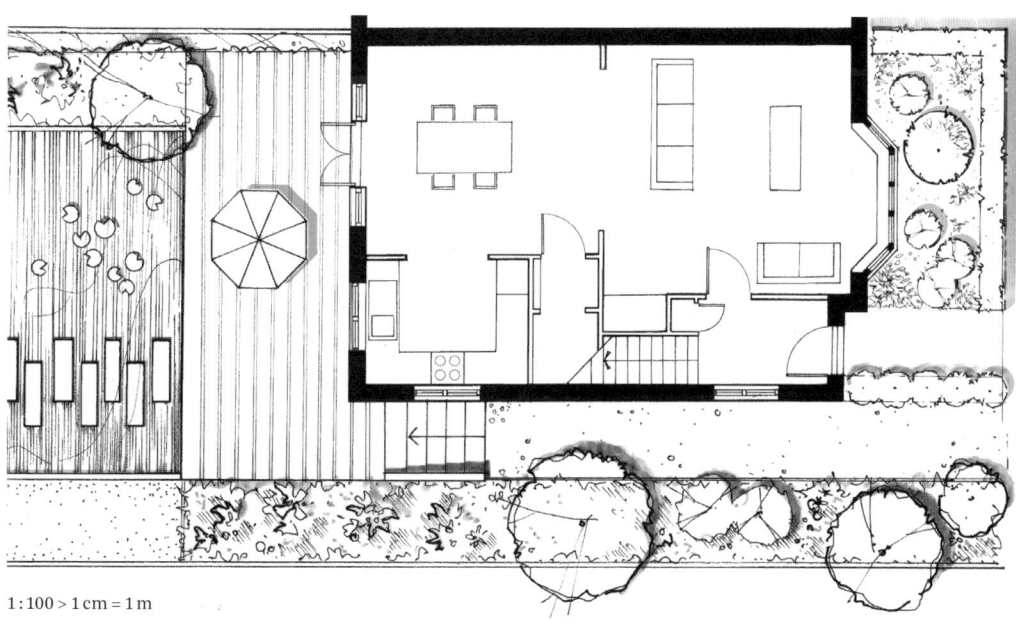

1:100 > 1 cm = 1 m

The plan view and the rendition of symbols

Floor plans and roof plans

Landscape architectural design usually relates to a greater context. Working in both urban and rural areas means that design drawings often involve buildings as well as topography. Roof plans show buildings from above and relay their overall shape and slopes, along with the overall building volumes and their location in relationship to the immediate surroundings or even to the general landscape. Floor plans are horizontal sections of a building, cutting through all major vertical elements of it. They reveal walls, doors, windows and most importantly, they communicate the internal organisation of a building. Any cut elements are highlighted graphically, usually through thicker line weights and infilled surfaces. These surfaces can be given a grey tonal value or made completely black. While the floor plan can be further enhanced with furniture to show uses and layouts, it is important not to let the internal details distract from the overall landscape design.

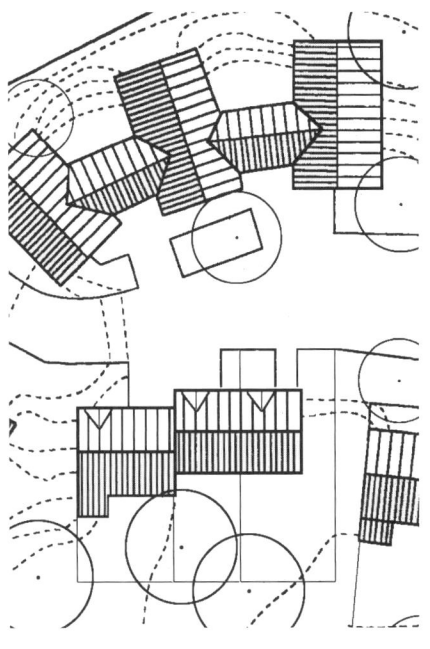

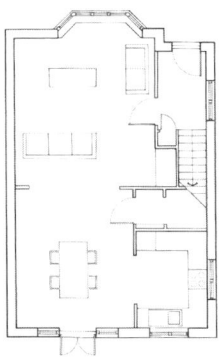 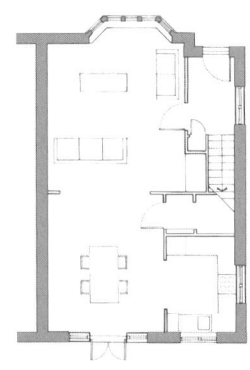 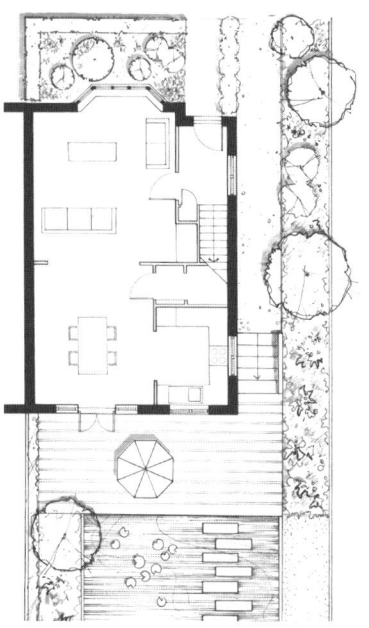

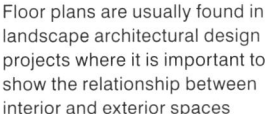

Floor plans are usually found in landscape architectural design projects where it is important to show the relationship between interior and exterior spaces

40

Buildings
Floor plans and roof plans
Roof plans and shadows

3

Roof plans and shadows

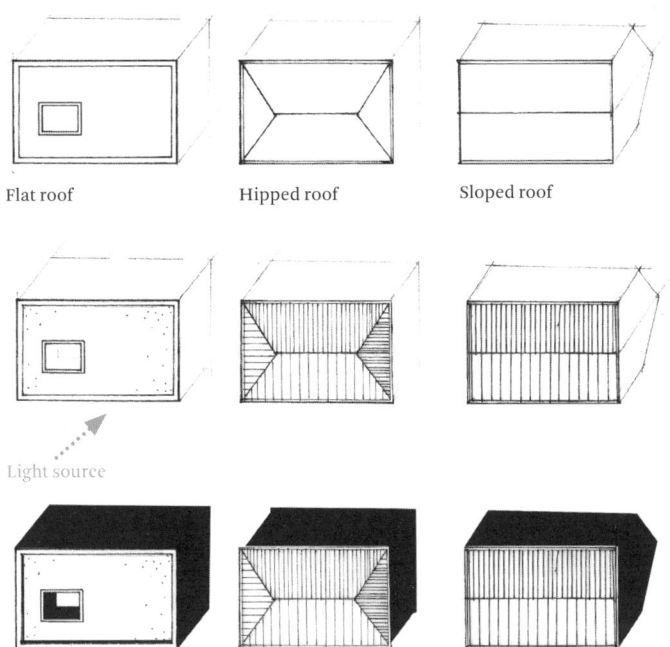

The scale usually will determine the level of detail with which a roof plan will be shown. In larger scales, roofs are shown with textures and shadows, giving their volumes three-dimensional effects. Smaller scale drawings often only require an outline showing the roof shape. In a figure ground plan, only the black footprint of a building form is needed.

After defining a light source in relationship to the roof plan, parallel lines are projected at the same angle from the corners of the shape. By joining up these lines and infilling the resulting shape with grey tone, or even black, the shadow area creates a strong contrast in tonal value which seemingly lifts the roof upwards. The height and mass of the building volume is emphasised, as is its relationship to the ground plane.

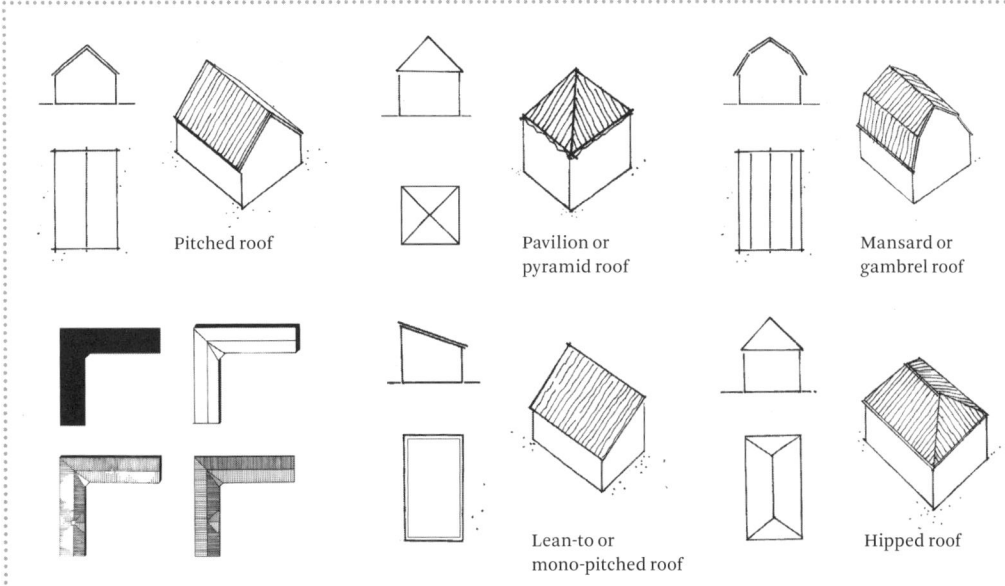

Since roofs are projected orthogonally in a plan, the result is an abstract and somewhat 'flat' representation, even if the roof in question has very distinctive slopes and angled parts. Here are some common roof types and their projections.

41

The plan view and the rendition of symbols

Drawing trees

Drawing trees is an inescapable part of landscape architectural graphics. Since design projects are concerned with designing outdoor spaces, they deal with a huge variety of natural vegetation. Every tree, shrub and plant is selected for its specific characteristics which relate to the site and for its value in an overall design concept. In a plan view, drawing trees involve different degrees of abstraction corresponding to the scale of the plan and the stage of the project. Tree symbols in plan can range in levels of detail between universality and particularity.

Universality
A universal symbol in plan and elevation tends to express an idea rather than a detailed tree type. These symbols can be interpreted as taken to mean any kind of tree or shrub.

Particularity
This drawing begins to show a particular type of tree and communicates its distinct form and characteristics, albeit still in an abstract and subjective way

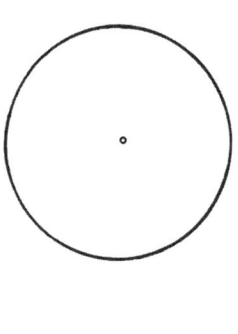

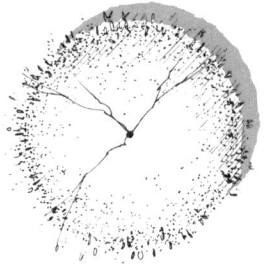

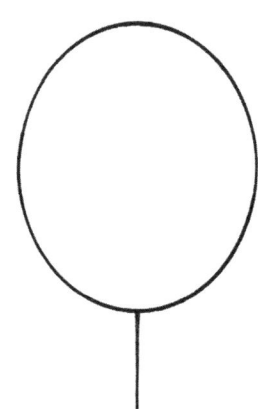

Trees and vegetation
Drawing trees
Tree symbols
Shade, shadows, and tonal values
Tree groups
Shrubs, hedges, and grass
Trimmed hedges and woody plants
Vegetation surfaces
Flowering plants
Sketching planting beds

3

In plan graphics, three symbols are always a graphic interpretation of a real thing. However, even though these examples are all idealised abstractions of a real tree type, they show how much can be communicated using only lines, textures and the circular form. How much detail goes into the graphic rendering of a tree depends on the time available, the viewer and the project phase. If it is important to convey a specific type of tree, key components such as leaf textures will become part of the graphic interpretation. Every landscape architect will develop his or her own graphic style for drawing trees. This process is similar to the development of handwriting, which also begins with standardised and universal lettering before evolving into a unique and personal graphic style over time.

Many different levels of abstraction lie between universality and particularity. Whether or not a tree or object is drawn precisely using many details or if its characteristics are merely suggested depends on the scale, the project phase, the audience and sometimes also the time available. Regardless of style and degree of abstraction, a presentation should always feature a graphic similarity and continuity of all different projections, such as plan, section, elevation.

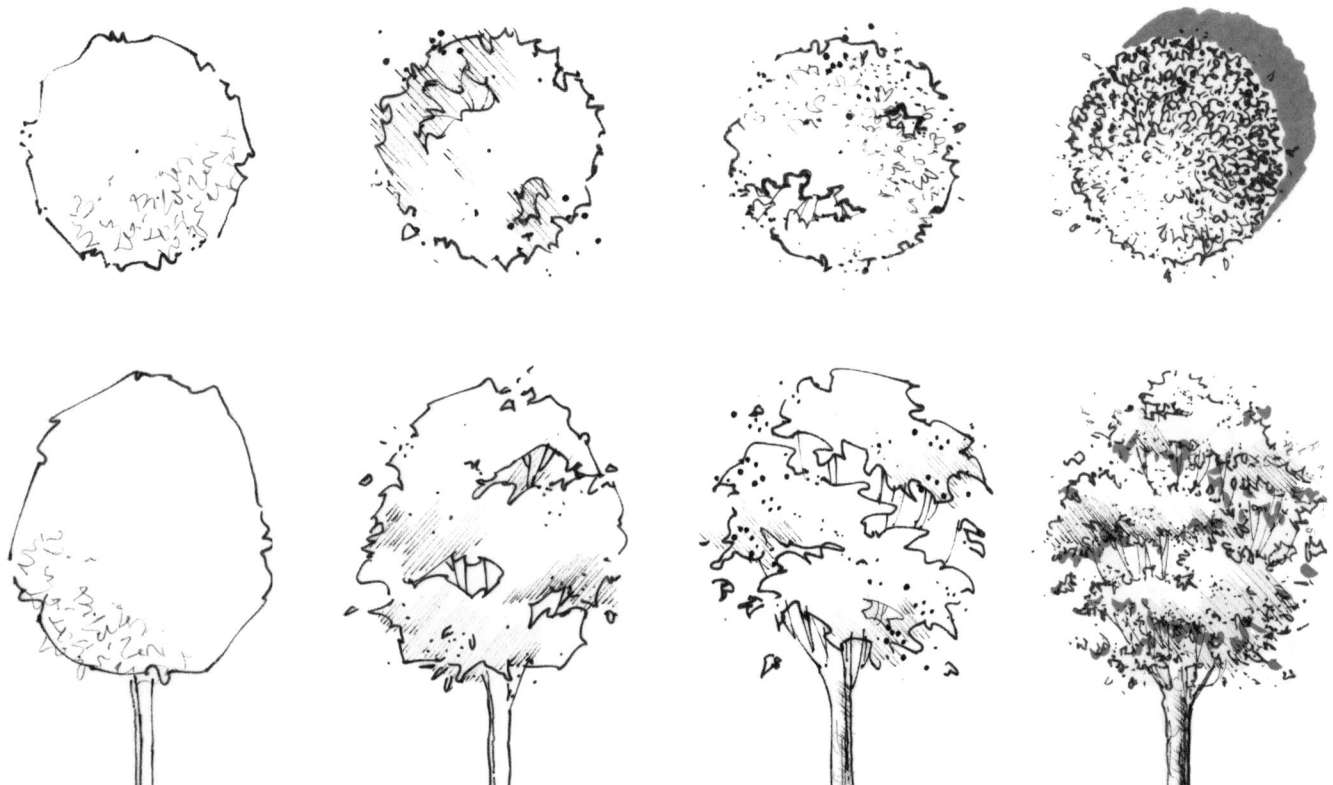

The plan view and the rendition of symbols

Drawing leaves and textures

In order to communicate a particular type of tree in both plan and elevation, it is good to begin by first looking at specific tree types and observing their leaves. These are first steps towards developing textures and graphic hatching in order to describe a tree's characteristics. While trees in plan might be circular, the hard geometric form doesn't communicate much about the tree's distinct foliage.

A tree's outline is usually irregular and composed of small parts. These can be communicated through lines and textures that have some relationship to the real tree leaves. It is a good idea to record leaves and their features in a sketchbook and then continually to abstract and reduce them until you arrive at a fast and legible tree symbol. Introducing a light source also helps to a sense of volume and mass to the shape.

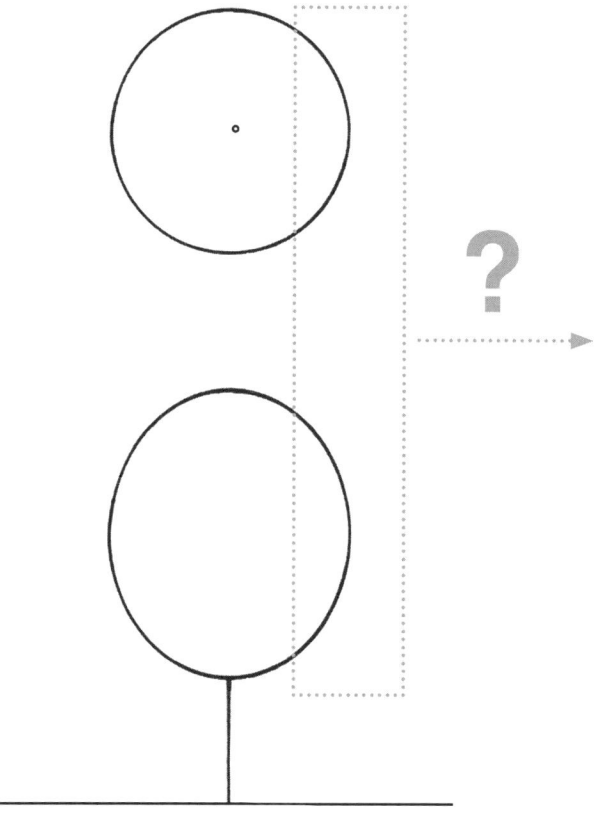

Tree shape in elevation
(see also page 116)

Trees and vegetation
Drawing trees
Tree symbols
Shade, shadows, and tonal values
Tree groups
Shrubs, hedges, and grass
Trimmed hedges and woody plants
Vegetation surfaces
Flowering plants
Sketching planting beds

3

First, a real leaf belonging to a specific tree must be observed and carefully sketched. In the process, its features are noted (long, pointy, upright, curly, dangling etc.).

The shape of the leaf can be made more and more abstract until it is reduced to a more graphic foliage texture

Using this texture to describe the shape of the tree helps break up the severity of the circle and suggest a specific tree type

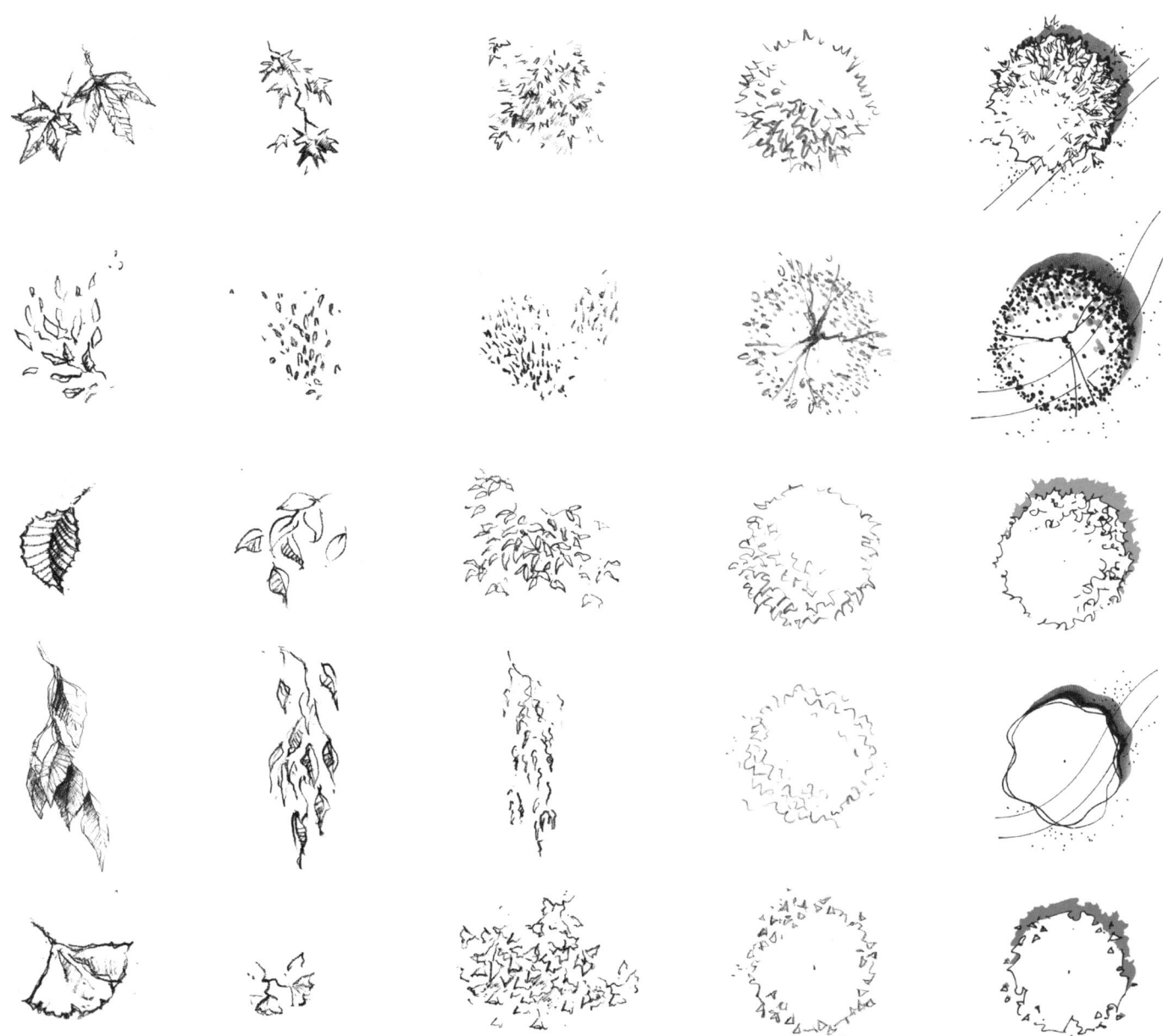

The plan view and the rendition of symbols

Drawing trees in plan

Trees in plan are usually drawn in an idealised form. Every tree symbol begins with a circle as its base. The middle of the circle suggests where the stem meets the ground plan and does not necessarily need special graphic emphasis. In a hand drawing, it is best to start with a pencil, using either a circle template or a compass to draw the circle form. As noted, the circle is a universal symbol for a tree, representing any tree type.

It is common to loosen up the edge of the circle using a textural hatching, especially since this aids identification and increases legibility. The tree symbol is recognised as an element of vegetation, rather than a hard geometric form. Small and irregular strokes and textures break up the pure circular base shape and, depending on the detail in the texture, can give some information about the tree, even if it is simply to distinguish a deciduous tree from an evergreen one.

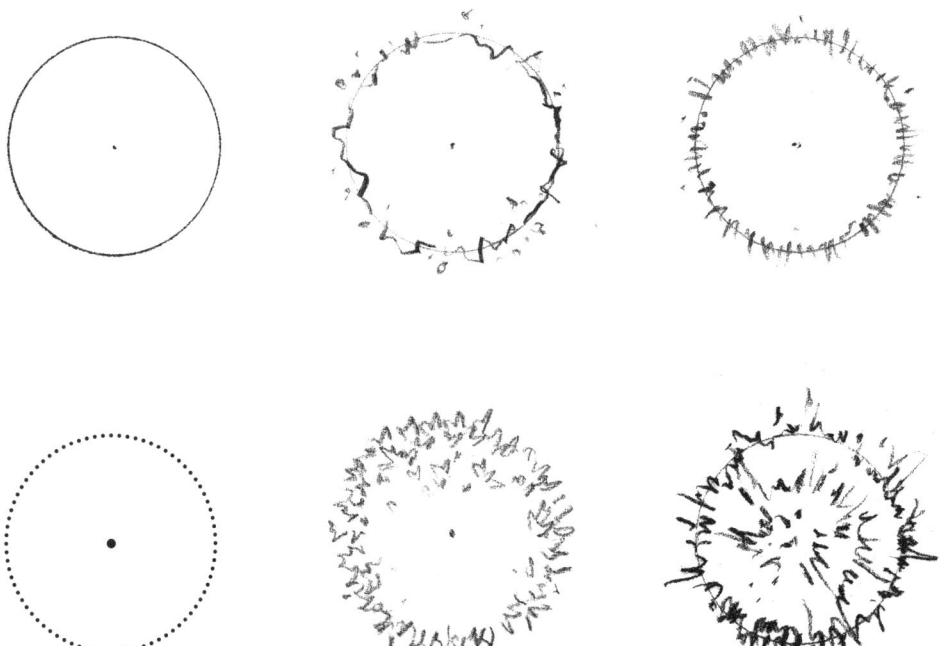

Trees and vegetation
Drawing trees
Tree symbols
Shade, shadows, and tonal values
Tree groups
Shrubs, hedges, and grass
Trimmed hedges and woody plants
Vegetation surfaces
Flowering plants
Sketching planting beds

3

With the help of a light source, it is possible to suggest the volume of the tree crown

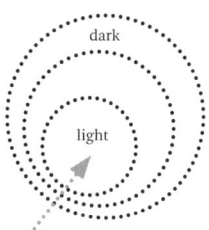

Light source from the southwest (afternoon sun)

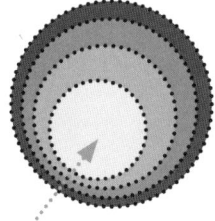

The area closest to the light source remains light

The area furthest away from the light source is darkest

By applying foliage texture in conjunction with a light source, tree crowns appear to become volumes, much like a sphere

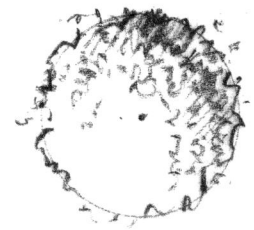

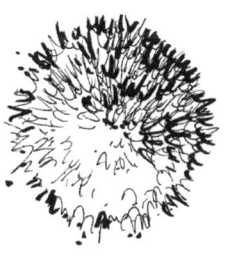

Varying degrees of foliage texture and gradient can help underscore the volume of the tree crown. Where the light source is closes to the tree, much can be left out.

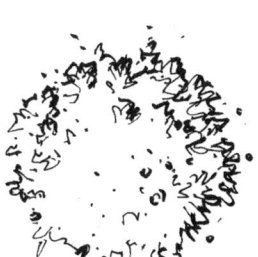

Adding more and increasingly dense foliage texture towards the area furthest away from the light source creates a gradient in grey tones. The resulting tree crown appears to be voluminous.

The plan view and the rendition of symbols

Tree symbols
Tree symbols should be easy and quick to draw. It is a good idea to try out as many as possible in a sketchbook in order to develop a more personal style and find the quickest and most comfortable symbols to draw for oneself.

Hand drawn trees can be loose and expressive, with a high degree of abstraction. These are useful for concept drawings and testing ideas.

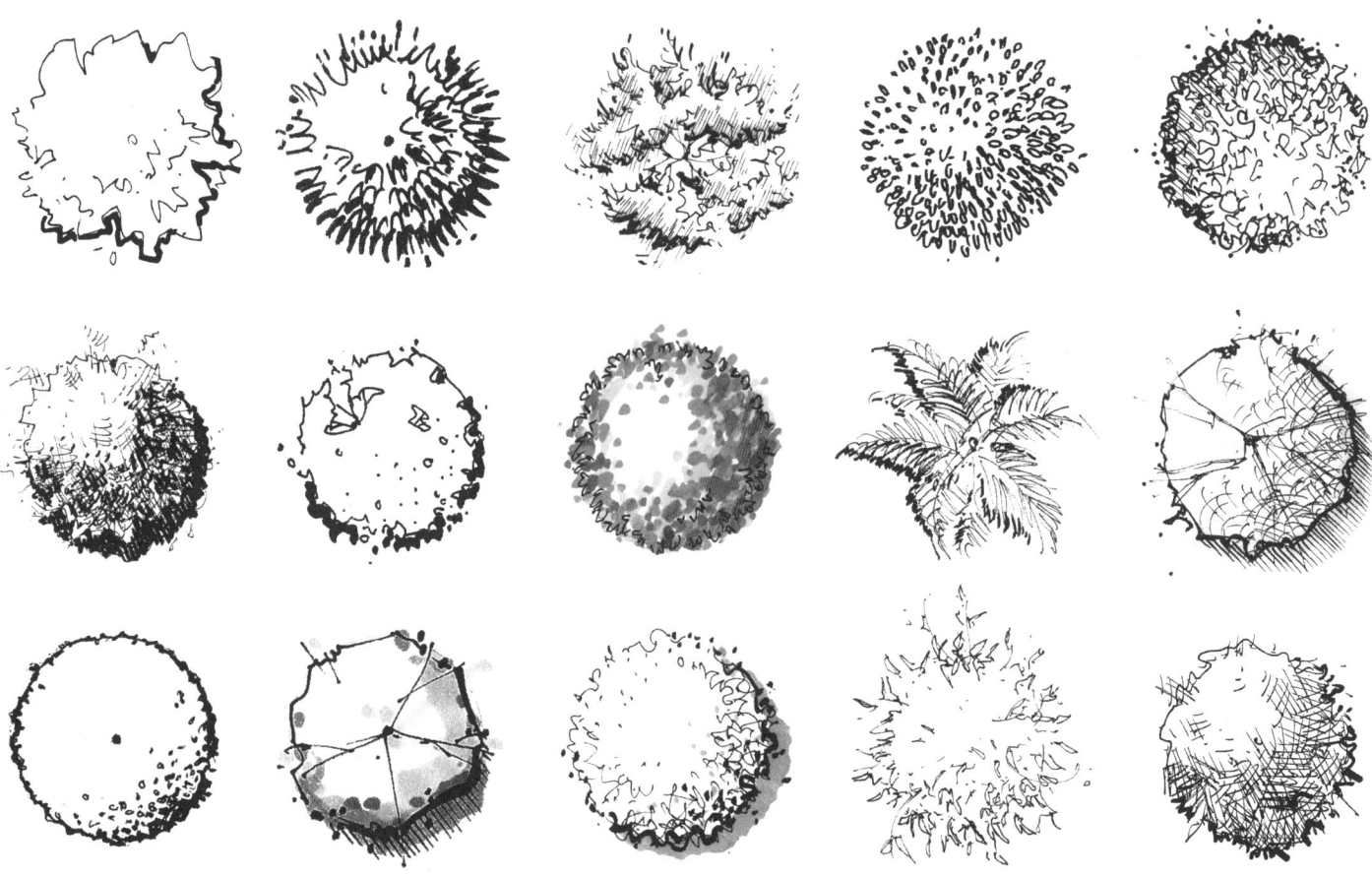

Trees and vegetation
Drawing trees
Tree symbols
Shade, shadows, and tonal values
Tree groups
Shrubs, hedges, and grass
Trimmed hedges and woody plants
Vegetation surfaces
Flowering plants
Sketching planting beds

3

These examples take longer to produce and have a more finished look to them. They also communicate more volume than abstract and quick sketch trees. When drawing foliage textures, it is important that the gradient doesn't become too dense on one side of the tree, while suddenly disappearing on the other. The gradient density should evolve gradually across the tree crown. The degree of darkness and detail in a tree will largely depend on its contextual surroundings in the plan.

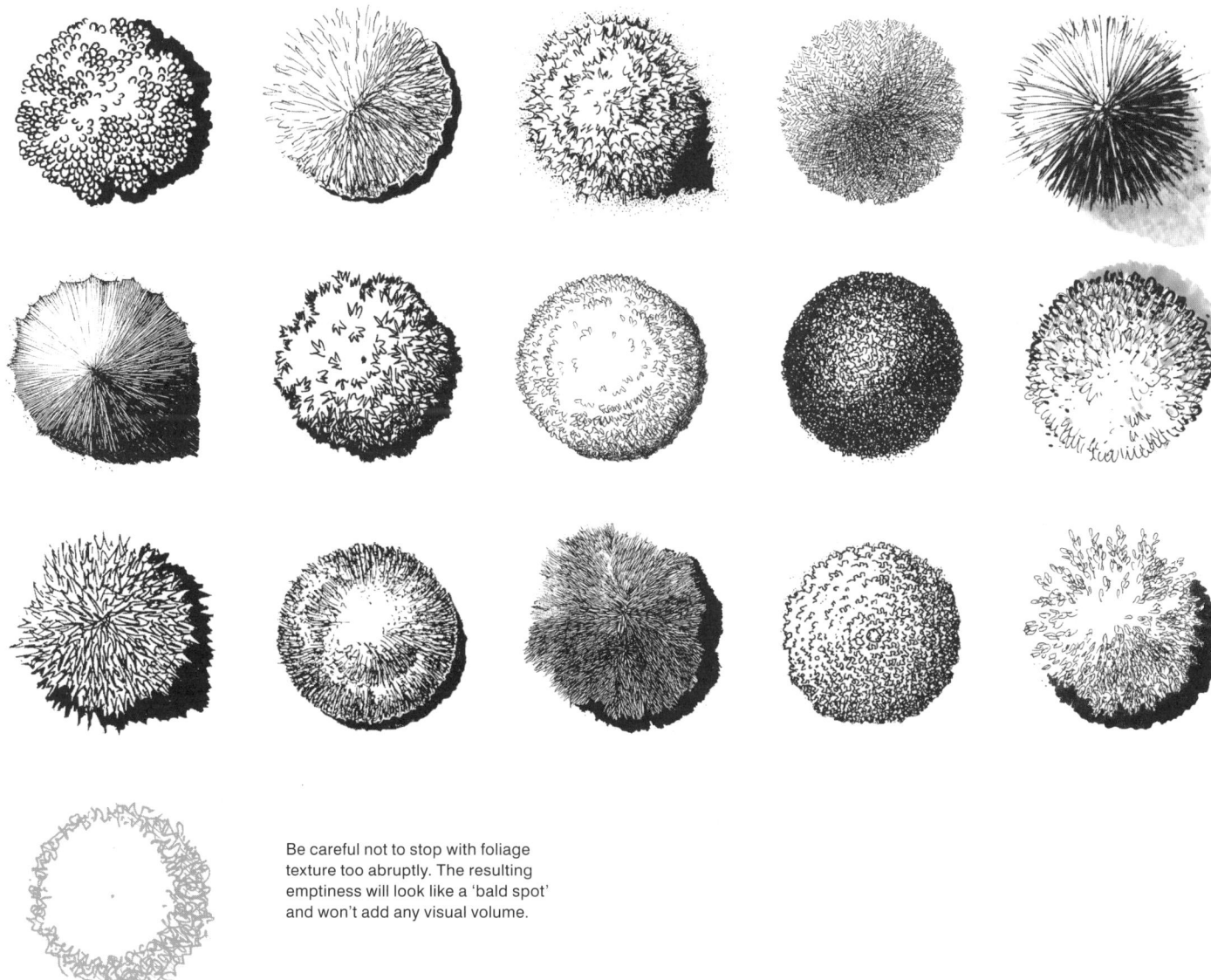

Be careful not to stop with foliage texture too abruptly. The resulting emptiness will look like a 'bald spot' and won't add any visual volume.

The plan view and the rendition of symbols

Quick trees
There is a wide range of interpretation when it comes to tree symbols. The quicker the drawing, the less foliage detail and the more gestures it will contain.

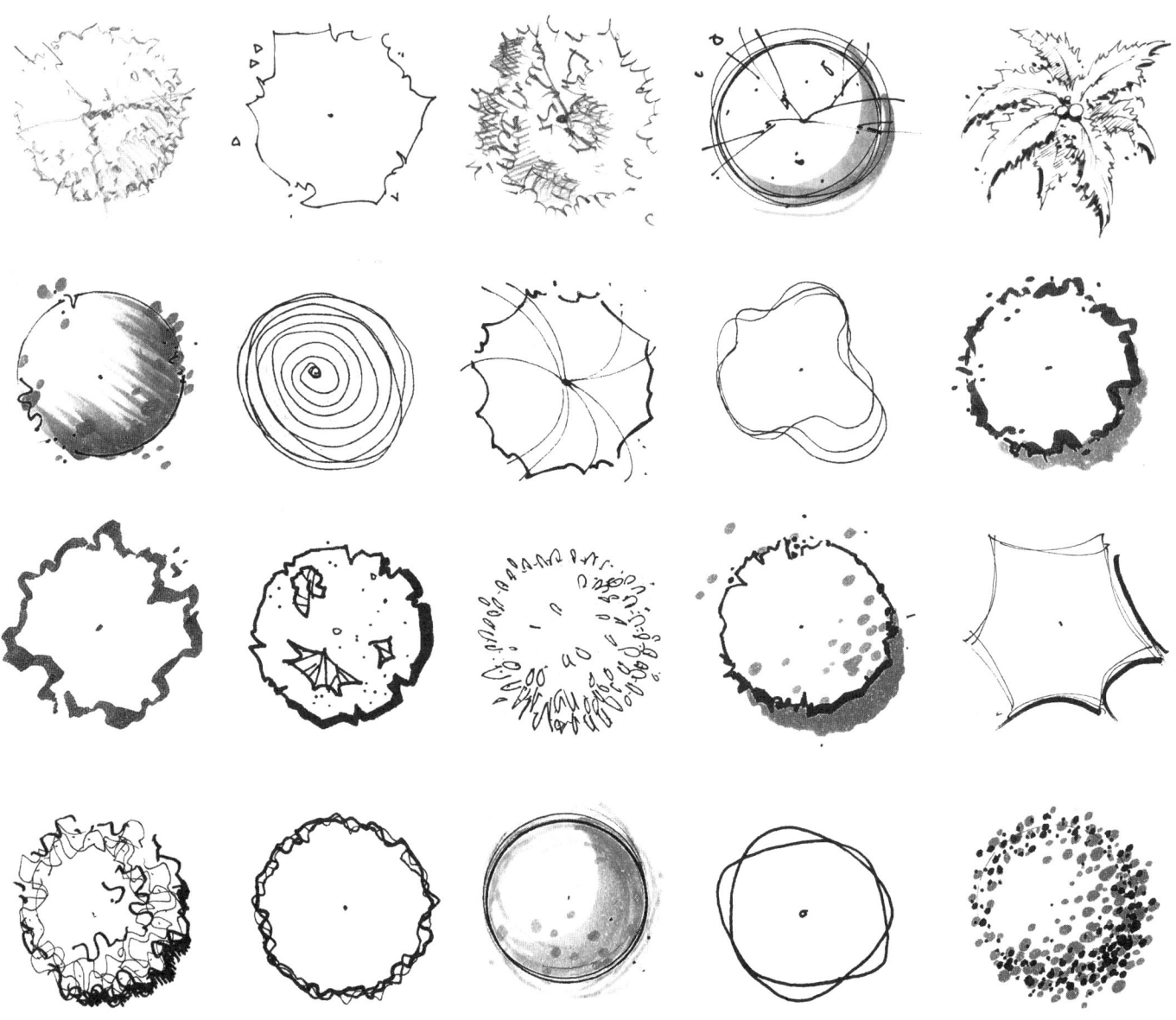

Quick deciduous trees

Trees and vegetation
Drawing trees
Tree symbols
Shade, shadows, and tonal values
Tree groups
Shrubs, hedges, and grass
Trimmed hedges and woody plants
Vegetation surfaces
Flowering plants
Sketching planting beds

3

Quick branch pattern and evergreen trees

The plan view and the rendition of symbols

Drawing trees with shadows
Shadows are not always necessary in a plan, however they can add a spatial effect to the drawing. The contrast provides visual impact and seemingly lifts the tree from the ground plane.

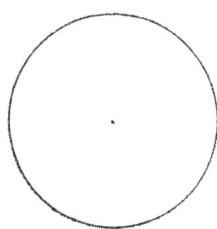

Circle outline as base form (in pencil)

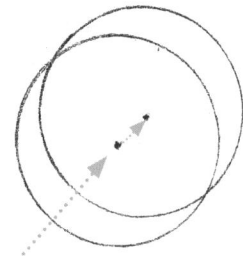

Light source (here, afternoon sun from southwest presuming north is upright). A second circle is drawn and moved in accordance with the direction of the light source.

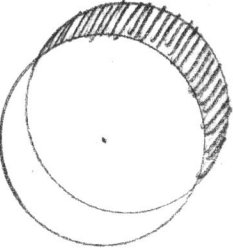

The resulting crescent shape is filled with a tonal value; the denser the shadow, the stronger the visual impact

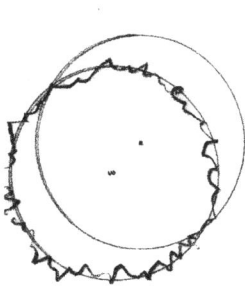

The shadow edge should be loosened up to correspond with the irregular outline of the tree

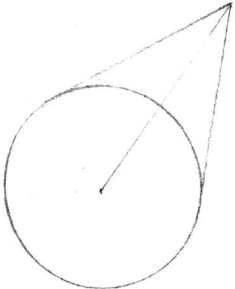

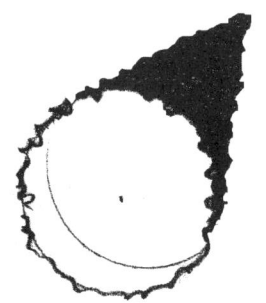

For coniferous or other trees with conical shapes, the shadow can have a longer, somewhat pointier shape

Trees and vegetation
Drawing trees
Tree symbols
Shade, shadows, and tonal values
Tree groups
Shrubs, hedges, and grass
Trimmed hedges and woody plants
Vegetation surfaces
Flowering plants
Sketching planting beds

3

Using shadows in a plan underscores the spatial volume of a tree and can also reveal information about the tree's characteristics.

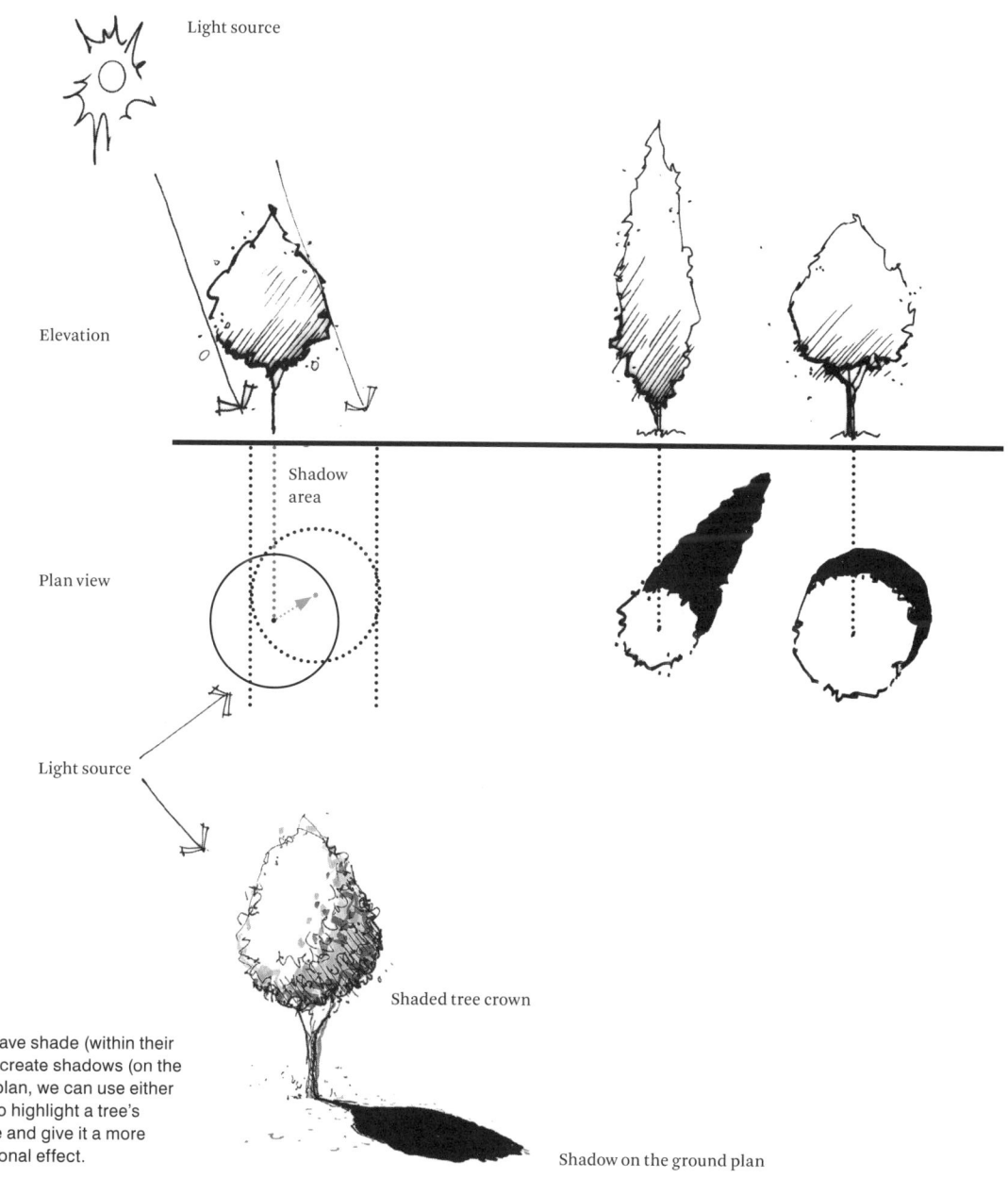

Tree crowns have shade (within their volumes) and create shadows (on the ground). In a plan, we can use either or even both to highlight a tree's spatial volume and give it a more three-dimensional effect.

The plan view and the rendition of symbols

Shade, shadows, and tonal values
Here is an example with several different graphic types of shadow around the trees. There are different hatchings and tones to be seen here. It is best not to mix solid black shadows and greytone shadows in the same drawing.

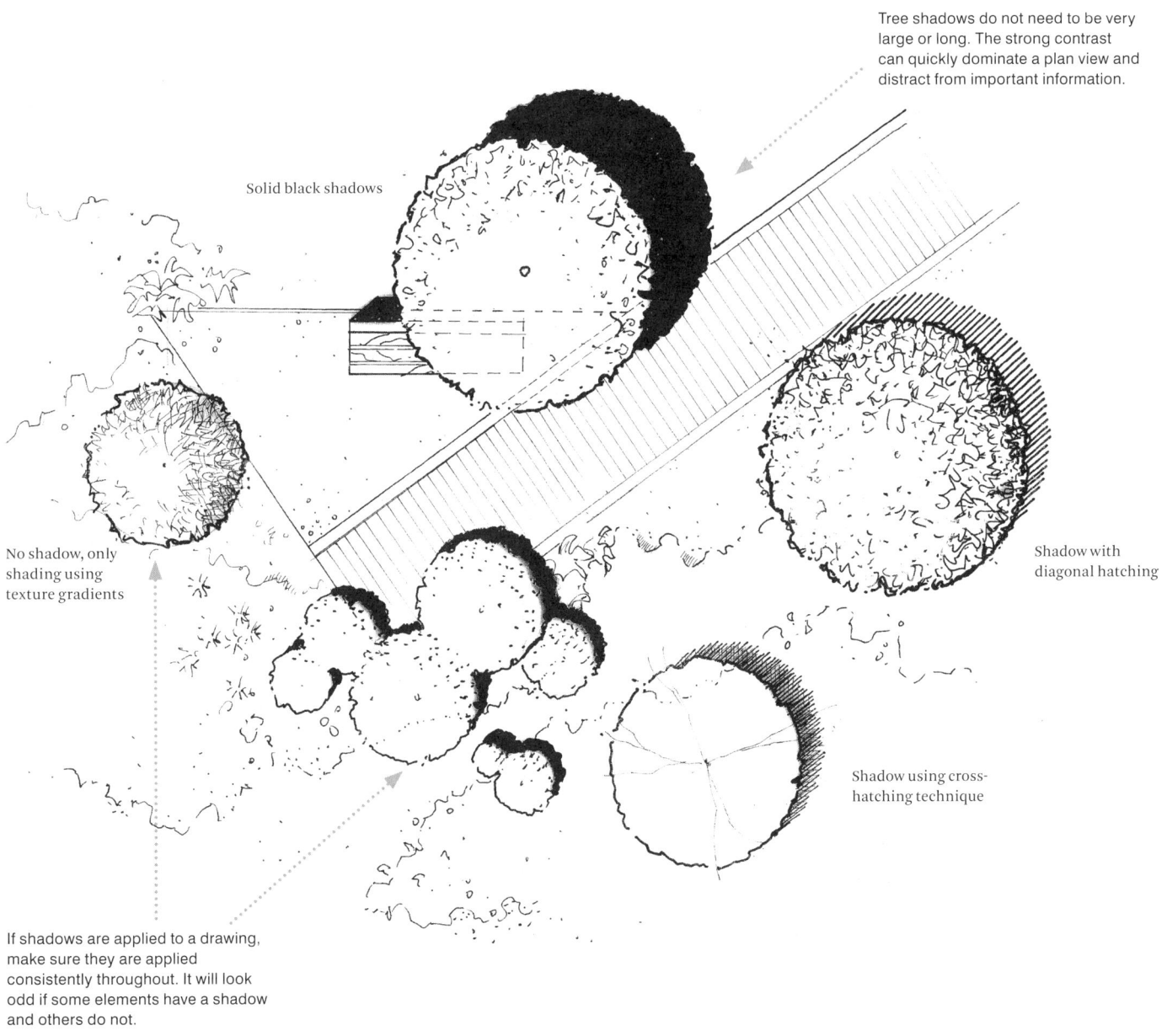

Solid black shadows

Tree shadows do not need to be very large or long. The strong contrast can quickly dominate a plan view and distract from important information.

No shadow, only shading using texture gradients

Shadow with diagonal hatching

Shadow using cross-hatching technique

If shadows are applied to a drawing, make sure they are applied consistently throughout. It will look odd if some elements have a shadow and others do not.

Trees and vegetation
Drawing trees
Tree symbols
Shade, shadows, and tonal values
Tree groups
Shrubs, hedges, and grass
Trimmed hedges and woody plants
Vegetation surfaces
Flowering plants
Sketching planting beds

3

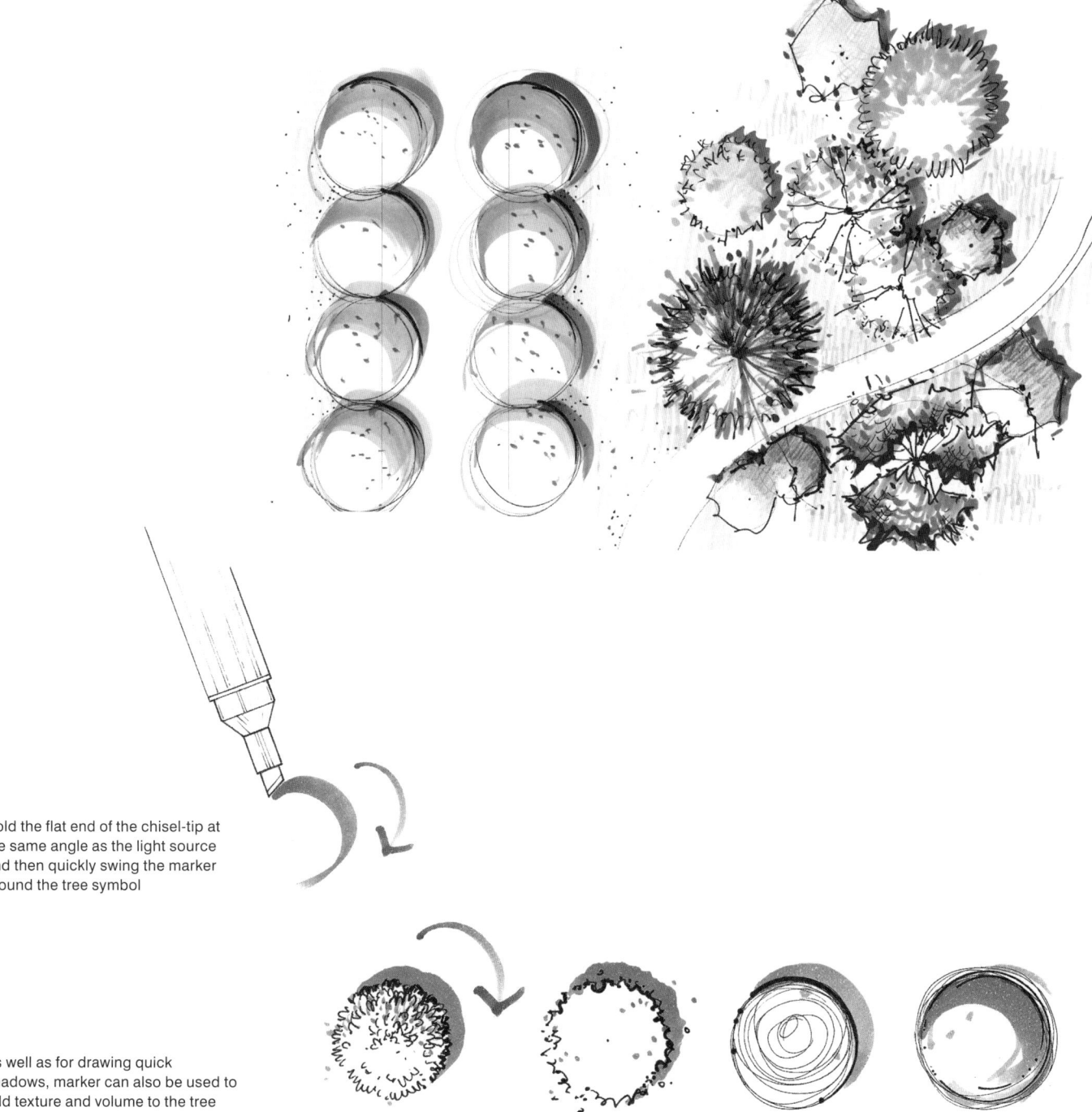

Hold the flat end of the chisel-tip at the same angle as the light source and then quickly swing the marker around the tree symbol

As well as for drawing quick shadows, marker can also be used to add texture and volume to the tree

The plan view and the rendition of symbols

Tree groups
In gardens or park situations, trees are often planted in groups. This means they do not have to be shown and rendered as separate entities. Individual circles are first drawn to represent each tree, followed by an overall outline which is applied to the whole ensemble. This graphically unifies all the trees into a larger, easily legible group.

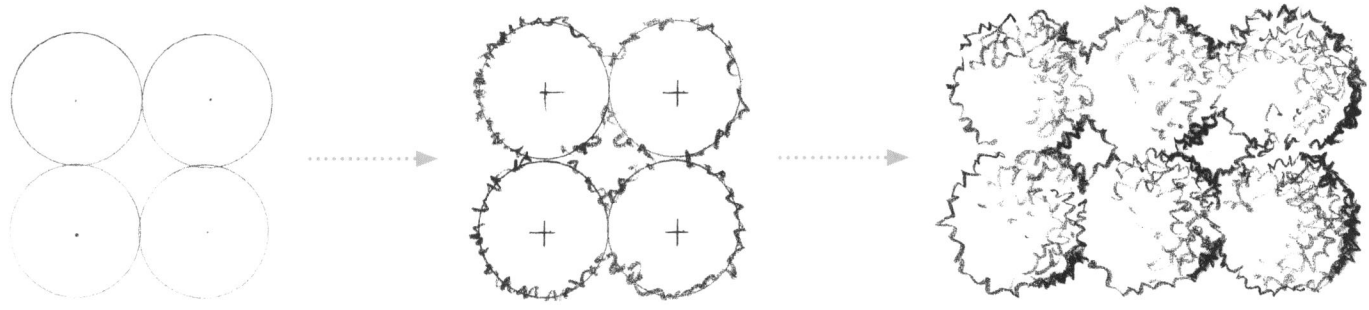

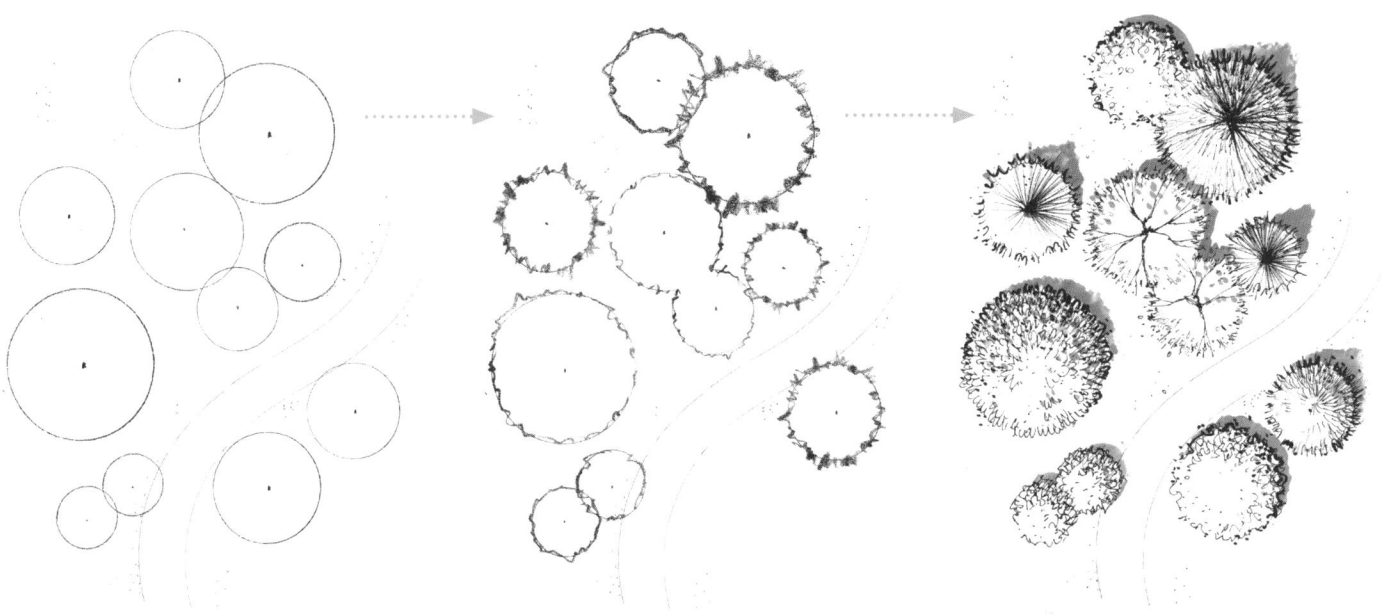

Trees and vegetation
Drawing trees
Tree symbols
Shade, shadows, and tonal values
Tree groups
Shrubs, hedges, and grass
Trimmed hedges and woody plants
Vegetation surfaces
Flowering plants
Sketching planting beds

3

Here is an example of a tree group, quickly drawn without the help of a circle template

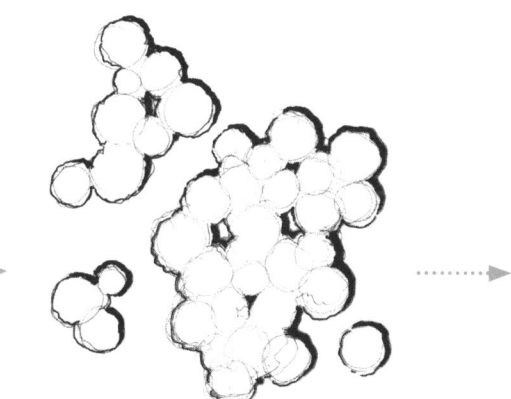

Preliminary sketch with differently sized circle forms, bound together by a common outline

Shadows are quickly drawn using a marker

Volumes and forms are suggested through greytones (for example through diagonal hatching)

Tree groups are a common part of landscape architecture projects and tend to carry more visual weight than any individual tree. It is important for the viewer to understand how the trees are arranged, how they relate to other elements in a design and how their placement define spaces.

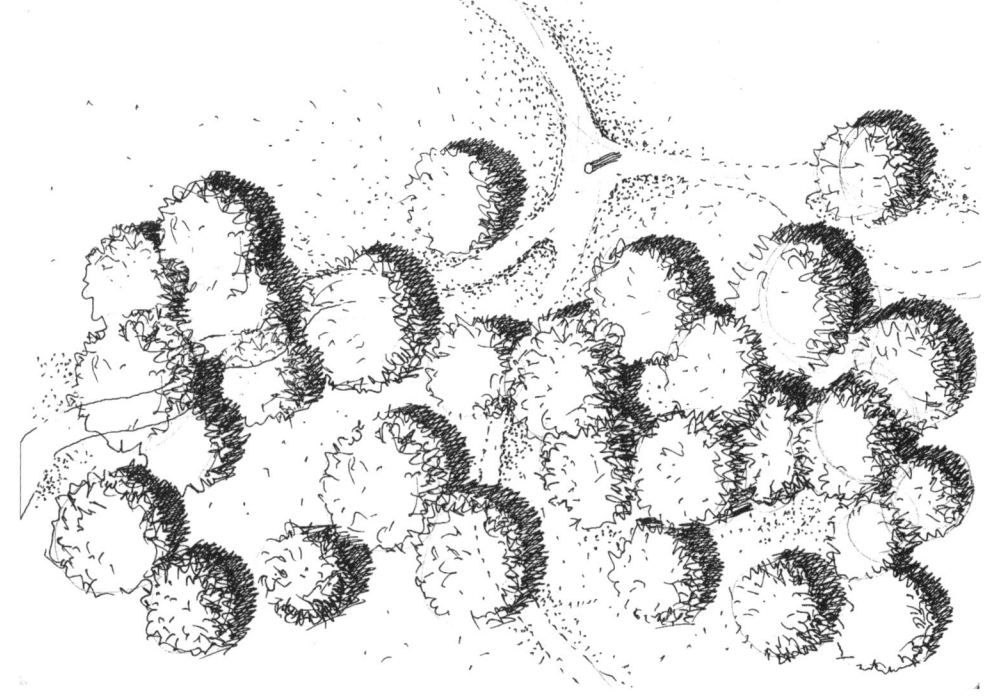

57

The plan view and the rendition of symbols

Trees with branch patterns
Branch pattern trees, sometimes also referred to as winter trees, can bring lively textures to a plan drawing. They do require extra time and practise to draw and can retain a diagrammatic effect. They are purely graphic, as they do not reveal anything about a tree type or its foliage.

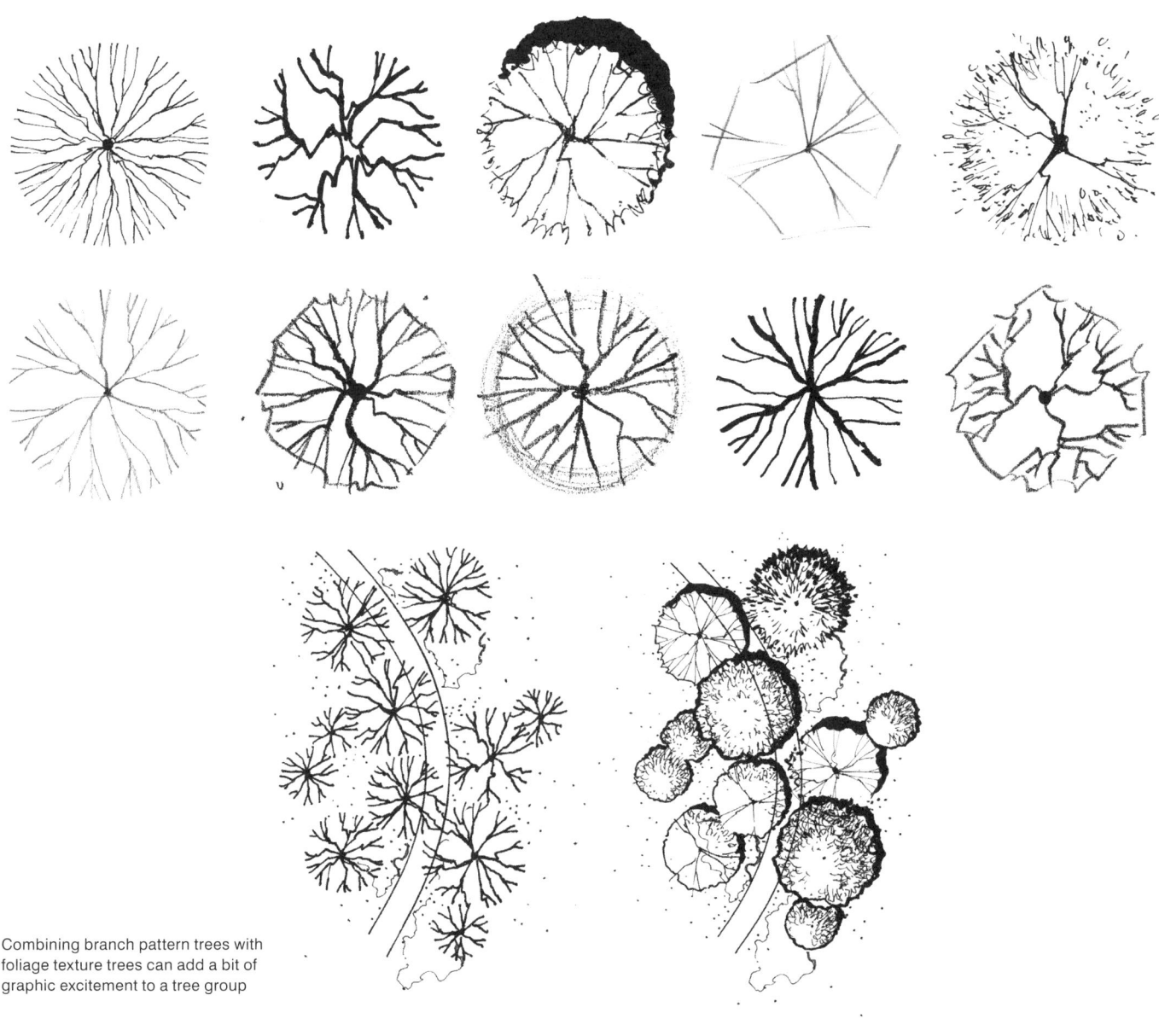

Combining branch pattern trees with foliage texture trees can add a bit of graphic excitement to a tree group

Trees and vegetation
Drawing trees
Tree symbols
Shade, shadows, and tonal values
Tree groups
Shrubs, hedges, and grass
Trimmed hedges and woody plants
Vegetation surfaces
Flowering plants
Sketching planting beds

3

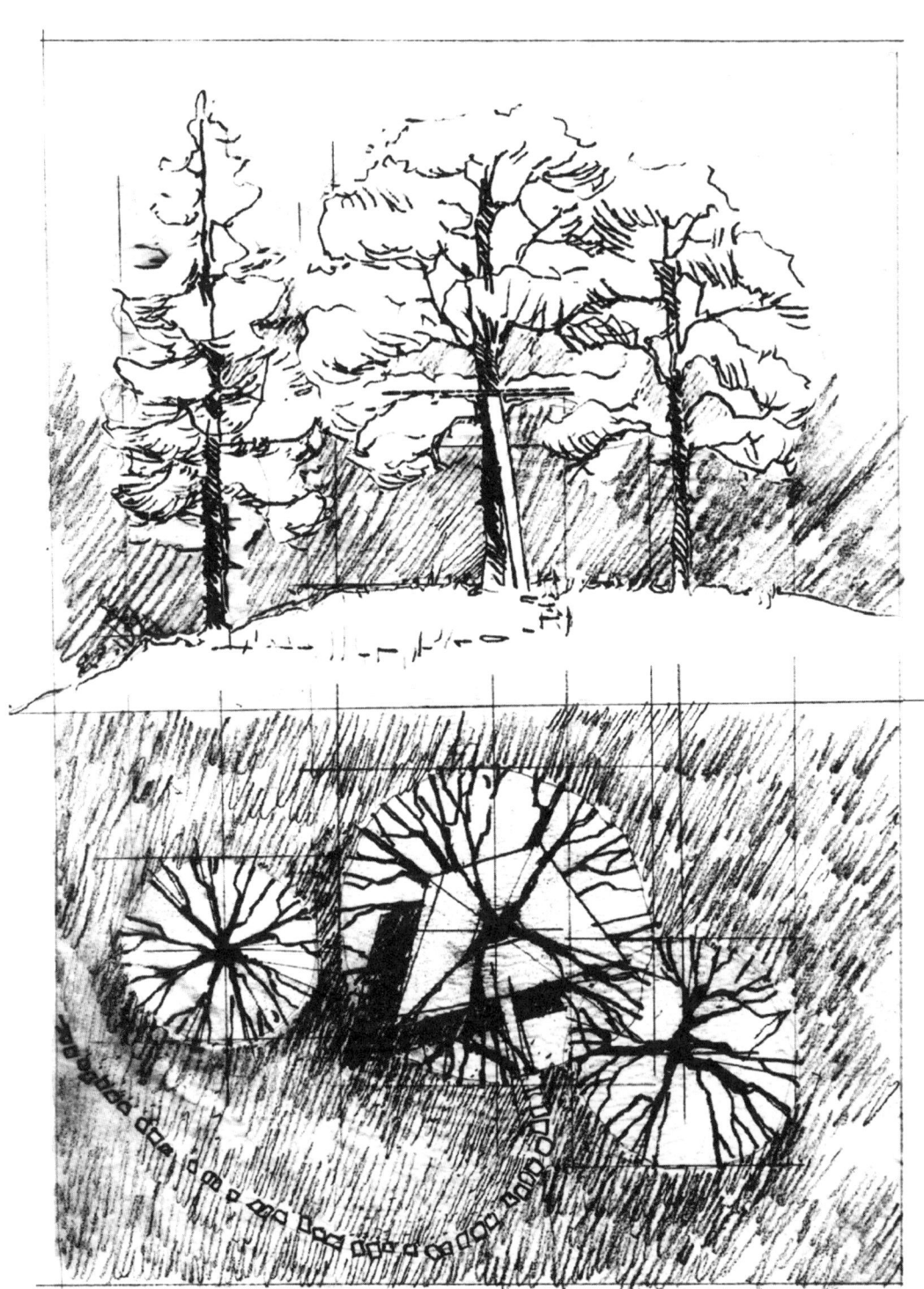

This example deliberately used branch pattern trees in the plan view, in order to reveal the tree platform within the crown

The plan view and the rendition of symbols

Reading tree groups and their positions

The more we zoom out of a drawing, the less important details and materials become. Instead, overall connections and relationships between the contents have to brought forward.

The way tree groups are drawn and arranged will communicate what is going on in the space. Allées, forests, orchards and naturalistic tree groups should be easily legible.

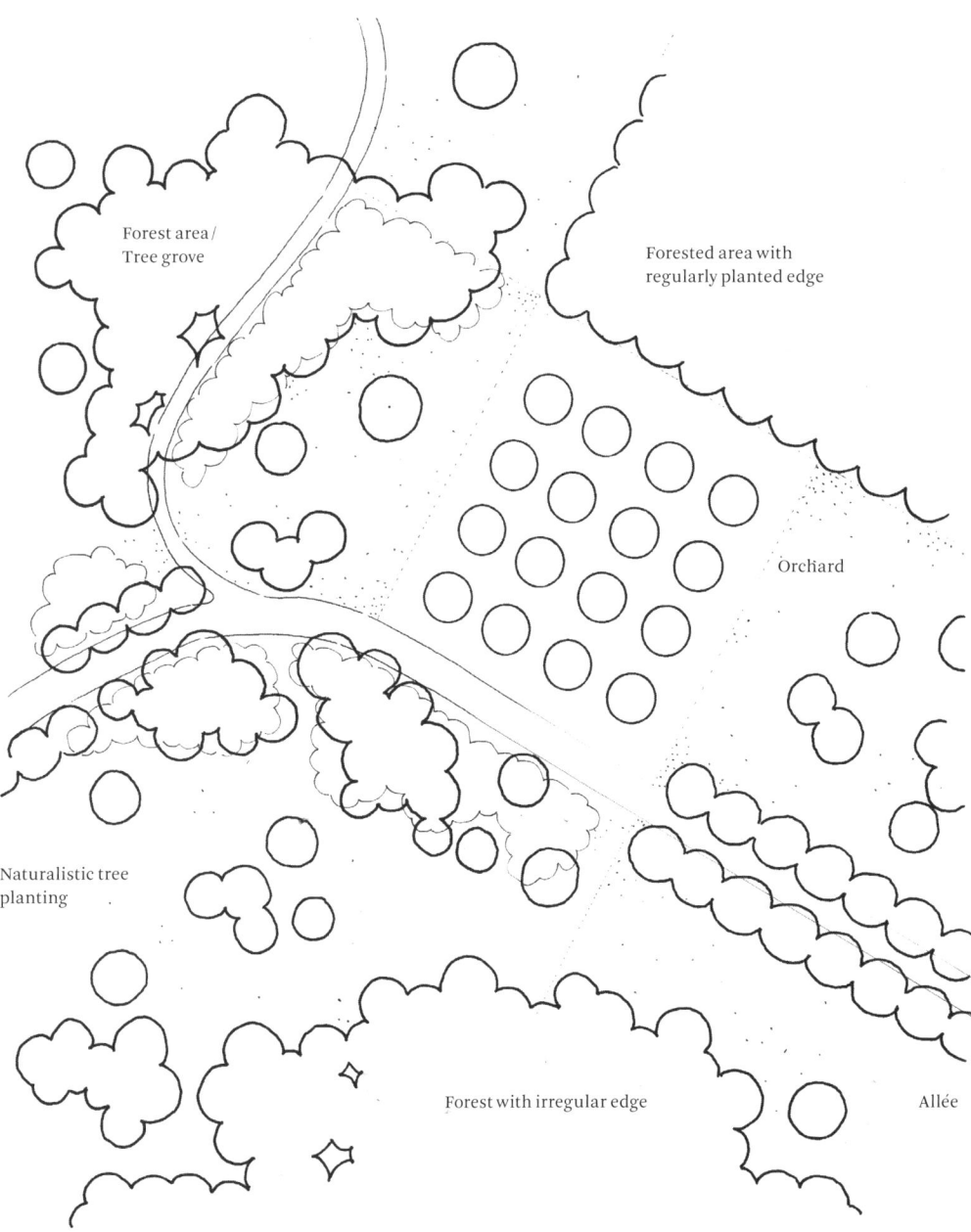

Trees and vegetation
Drawing trees
Tree symbols
Shade, shadows, and tonal values
Tree groups
Shrubs, hedges, and grass
Trimmed hedges and woody plants
Vegetation surfaces
Flowering plants
Sketching planting beds

3

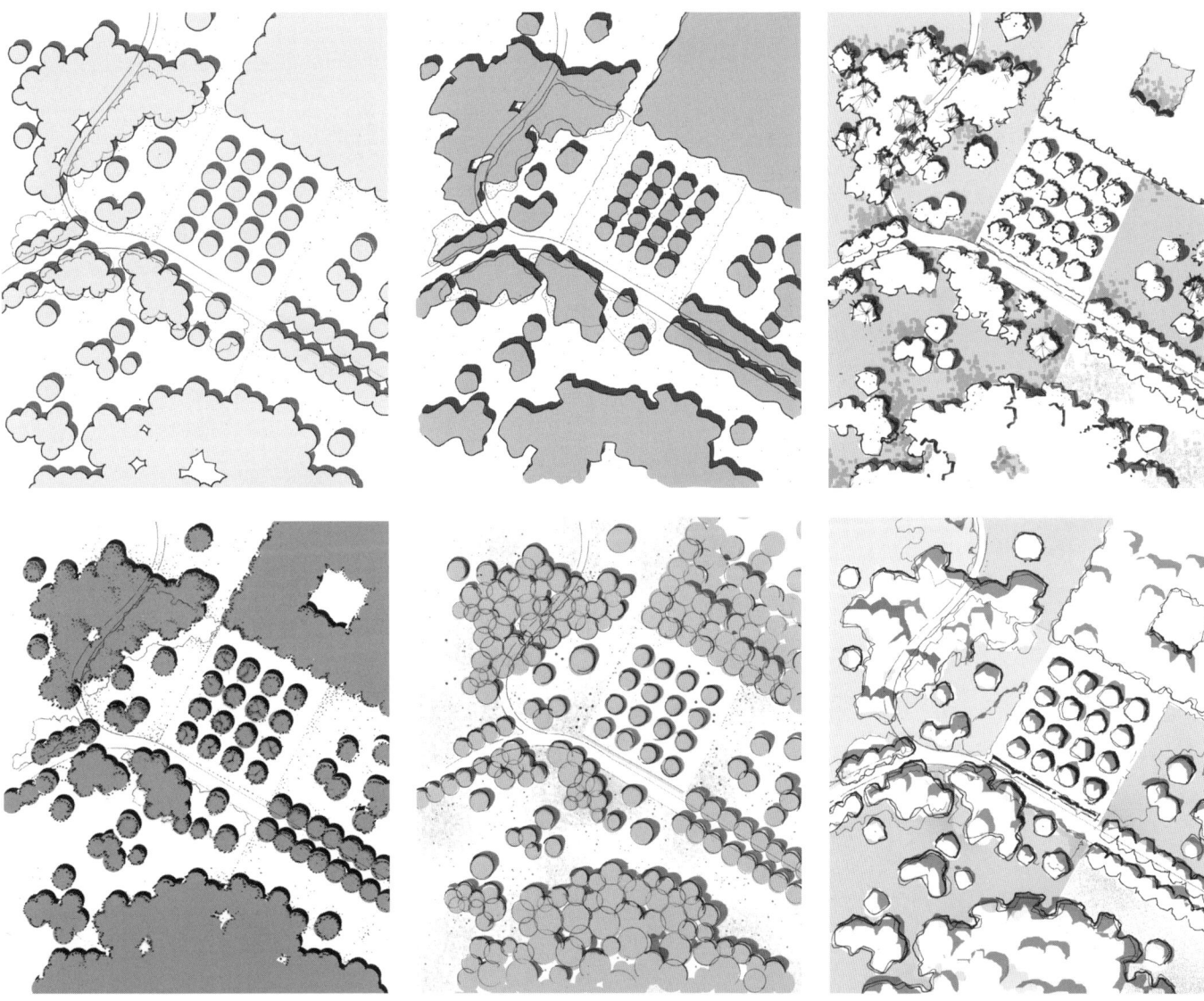

Here are some variations using grey tones. Dark trees on an lighter ground plane are always easily legible. Shadows may not always be necessary but they can add three-dimensionality to the forms.

If tree groups are shown lighter than the ground plane, they will need shadows. These will allow them to visually emerge upwards and be legible as volumes.

The plan view and the rendition of symbols

Tree groups: Examples
Here are some good examples showing the range of possibilities on offer when drawing tree groups. Some contain lots of graphic detailing with shade and shadows, while others remain abstracted.

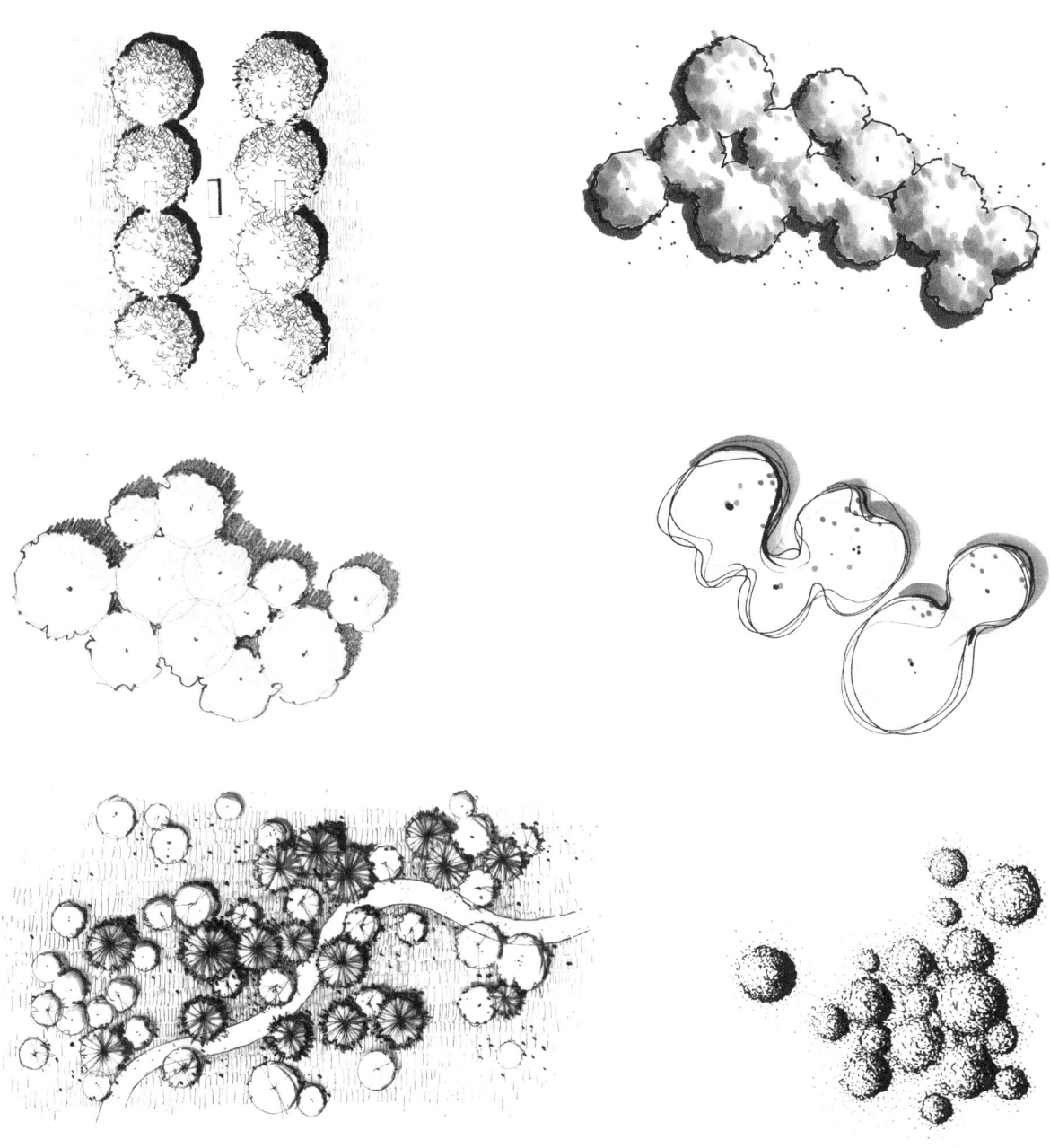

Trees and vegetation
Drawing trees
Tree symbols
Shade, shadows, and tonal values
Tree groups
Shrubs, hedges, and grass
Trimmed hedges and woody plants
Vegetation surfaces
Flowering plants
Sketching planting beds

3

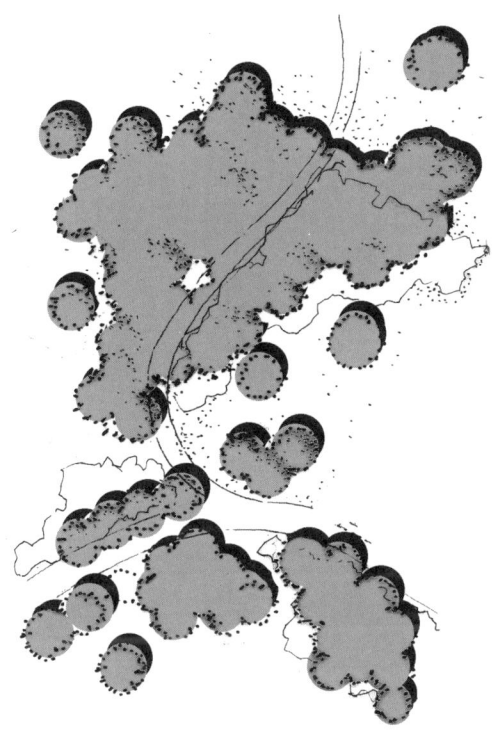
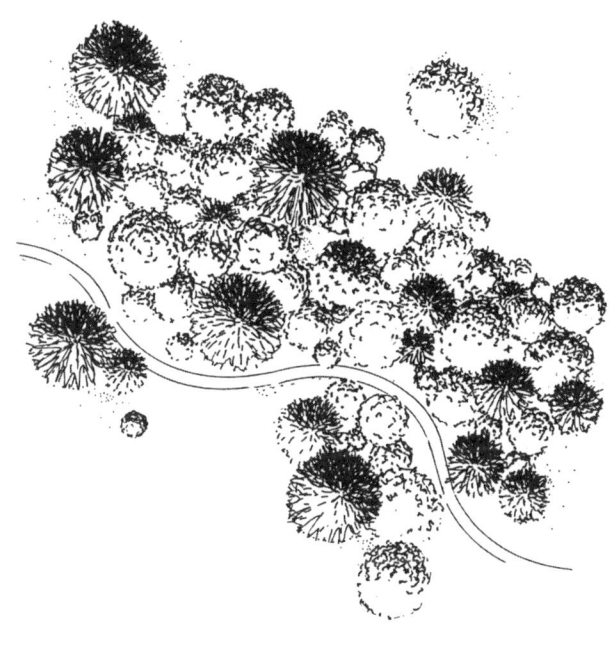
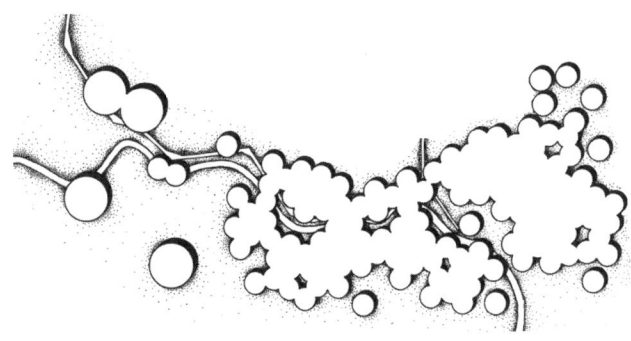

The plan view and the rendition of symbols

Shrubs, hedges, and grasses
Drawing shrub groups is similar to the method used for trees. Single plants are first drawn individually, then unified with an outline or irregular texture.

The texture style can refer to the specific characteristics of the plant itself or it can remain highly abstracted.

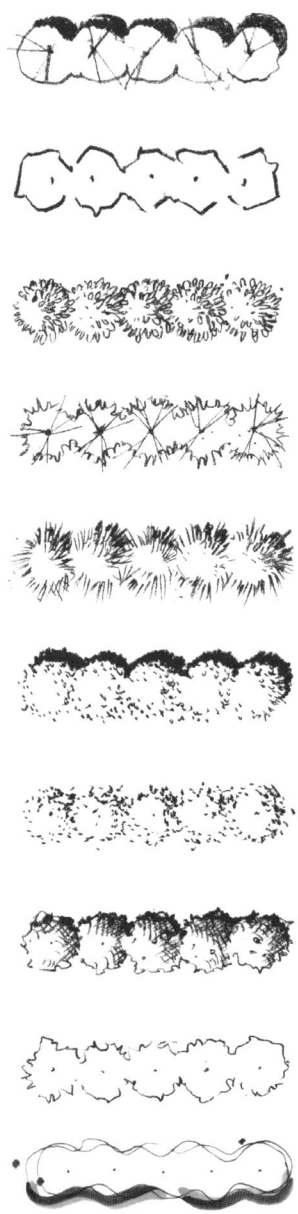

Shown here are several examples showing how shrubs, hedges and woody plants can be drawn in a plan. The less detail and small-scale texture, the most abstract the effect will be.

Trees and vegetation
Drawing trees
Tree symbols
Shade, shadows, and tonal values
Tree groups
Shrubs, hedges, and grass
Trimmed hedges and woody plants
Vegetation surfaces
Flowering plants
Sketching planting beds

3

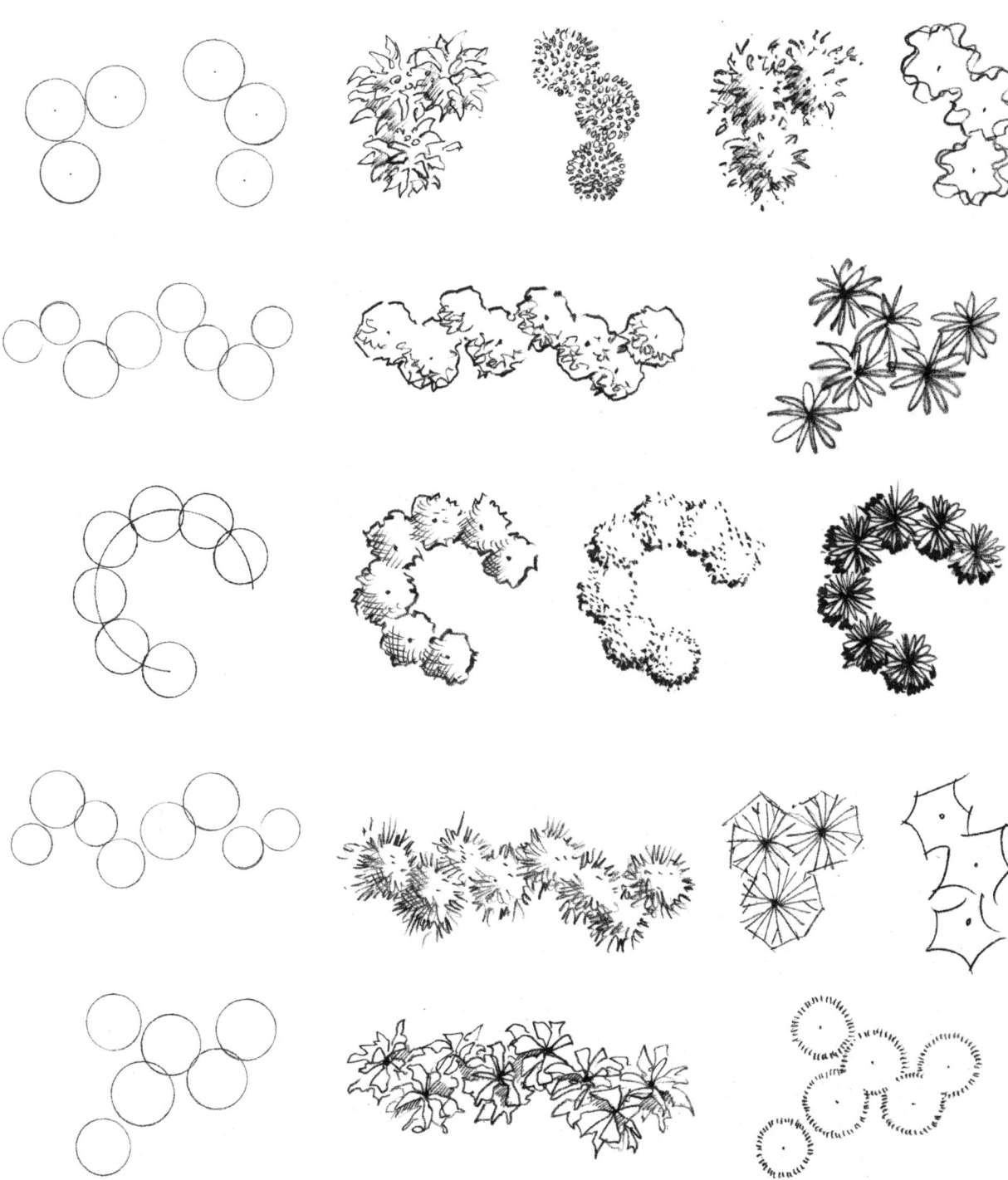

65

The plan view and the rendition of symbols

Trimmed hedges, woody plants, and topiary
Drawing clipped or trimmed hedges begins with a hard geometric base form. This is then given a looser and more irregular outline, whilst still maintaining a relatively straight edge.

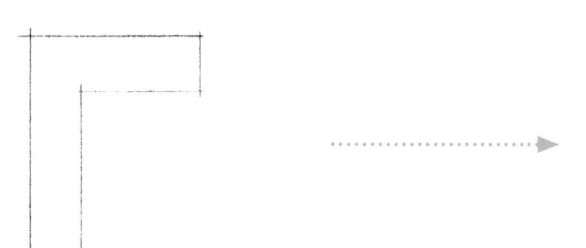

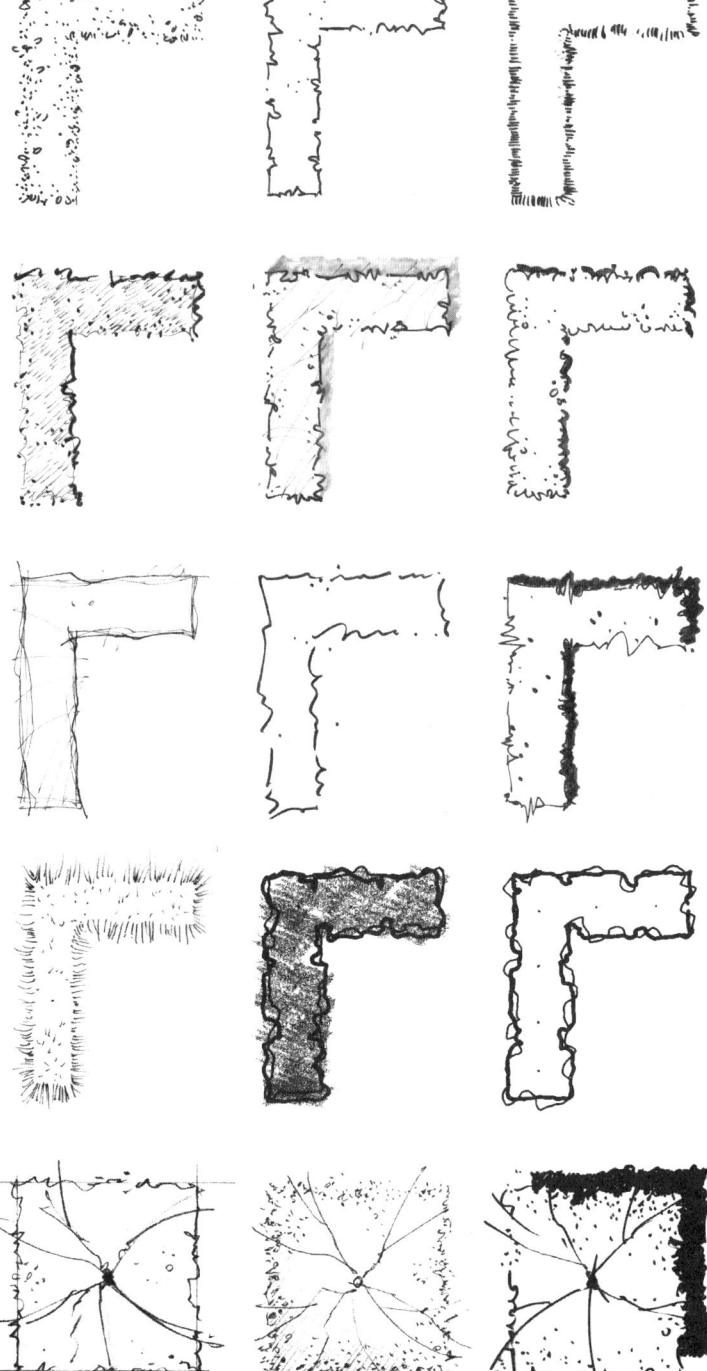

Drawing other topiary, clipped shrubs or pleached trees also begins with a sketch of the overall geometric shape. This outline must then be loosened up with an irregular texture to indicate to the viewer that the form is indeed vegetation.

Trees and vegetation
Drawing trees
Tree symbols
Shade, shadows, and tonal values
Tree groups
Shrubs, hedges, and grass
Trimmed hedges and woody plants
Vegetation surfaces
Flowering plants
Sketching planting beds

3

The faster the sketch, the less detailed the textures

The plan view and the rendition of symbols

Trimmed hedges, woody plants, and topiary
There are many ways to communicate clipped hedges and woody plants in a plan. Here are some more useful examples.

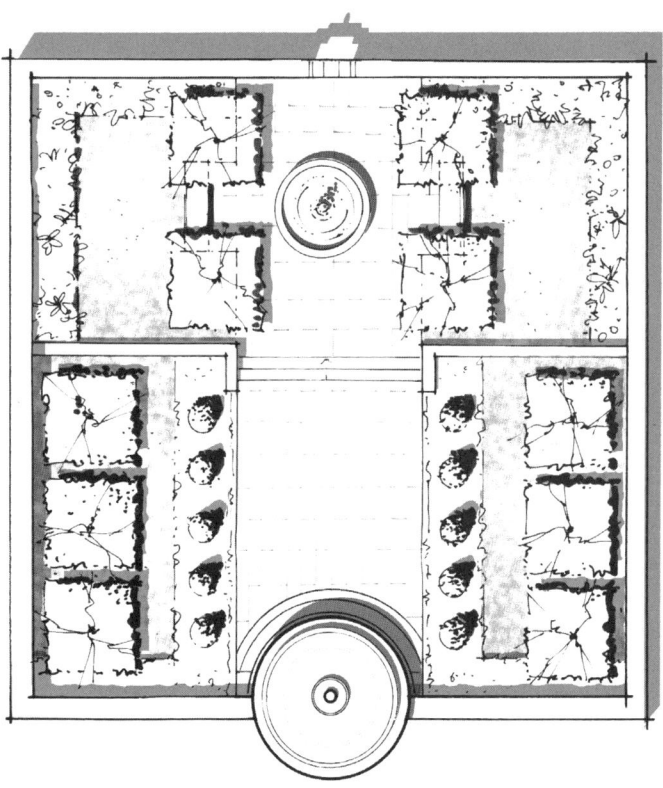

3

Trees and vegetation
Drawing trees
Tree symbols
Shade, shadows, and tonal values
Tree groups
Shrubs, hedges, and grass
Trimmed hedges, and woody plants
Vegetation surfaces
Flowering plants
Sketching planting beds

Although topiary is often associated with traditional, formal and historic gardens, it is still a part of contemporary garden design.

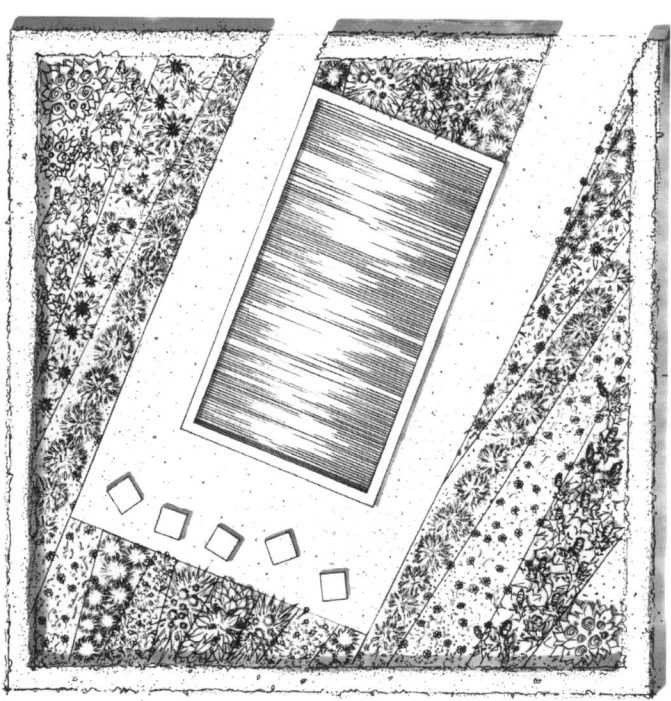
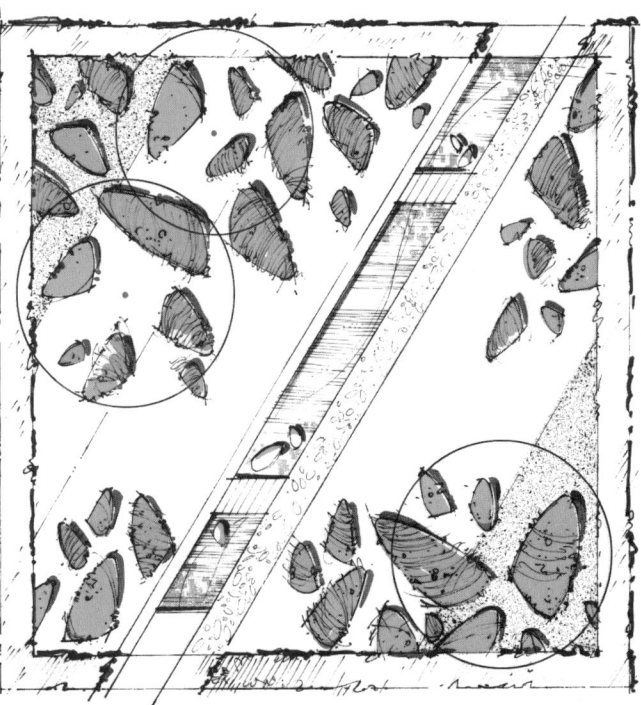

The plan view and the rendition of symbols

Ground cover and vegetation textures

Here are several examples of how ground covers can be drawn in a plan. Remember to ensure that there is enough contrast between the different textures when they are drawn next to each other.

These contrasts can occur by varying light and dark areas, as well as through differing grades of textural detail. The more similar two textures are, for example in grey scale, the less legible each becomes.

Grass, lawn and meadow areas

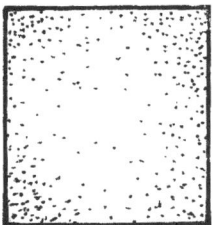 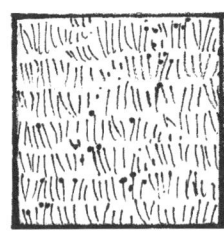 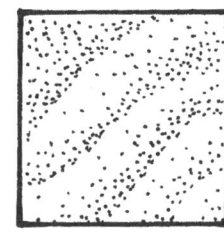 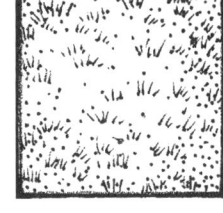

Using dots to indicate grass is simple and quick

Ground cover and low vegetation areas

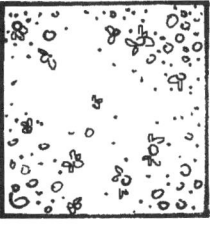   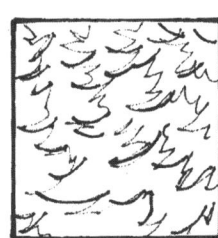

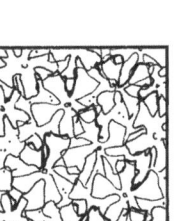 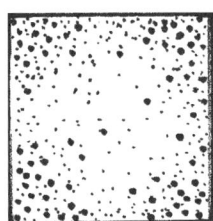 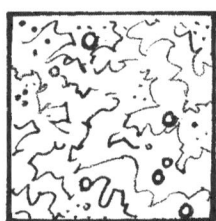 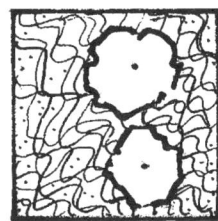

Trees and vegetation
Drawing trees
Tree symbols
Shade, shadows, and tonal values
Tree groups
Shrubs, hedges, and grass
Trimmed hedges and woody plants
Vegetation surfaces
Flowering plants
Sketching planting beds

3

The plan view and the rendition of symbols

Flowering plants

Even if flower beds are planted using a systematic layout, they should not be drawn that way. The irregular placement of blossoms and flowers across the planted area, with some areas more dense than others, will result in a much livelier and more realistic effect within the drawing.

Draw and distribute flower symbols irregularly within a planting bed

Here is a variety of symbols for blossoms and flowering plants

Trees and vegetation
Drawing trees
Tree symbols
Shade, shadows, and tonal values
Tree groups
Shrubs, hedges, and grass
Trimmed hedges and woody plants
Vegetation surfaces
Flowering plants
Sketching planting beds

Sometimes it is necessary to clearly indicate a specific type of vegetation within a planting plan. Key features of the plant, such as its flowers and leaves will have to be graphically suggested in the drawing. Here are two examples which display the process of abstraction from plant image to plan graphic.

Allium has long pointy leaves and large round blossoms

Hostas have large heart-shaped, densely-overlapping leaves with delicate, long flowering heads

These characteristics are then graphically interpreted into a plan view

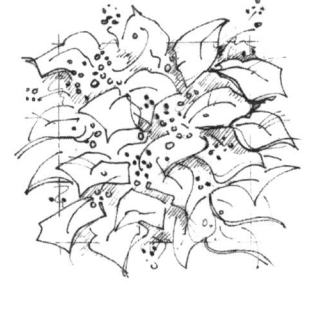

The degree of abstraction can be varied according to the time available and the viewer's ability to interpret the drawing

The plan view and the rendition of symbols

Drawing flowering plants and blossoms
Similar to drawing ground covers, it is important to achieve easy legibility when putting together the many textures of a flowering planting bed.

Differing vegetation textures need to contrast each other in density and detail, ensuring that each is recognisable to the viewer's eye.

When drawing potted plants, there are different options. If the plant overhangs the vessel, it can be indicated through a dotted outline underneath the plant. If the plant is slightly smaller than the pot, the round edge will be visible along with the plant.

Trees and vegetation
Drawing trees
Tree symbols
Shade, shadows, and tonal values
Tree groups
Shrubs, hedges, and grass
Trimmed hedges and woody plants
Vegetation surfaces
Flowering plants
Sketching planting beds

3

The plan view and the rendition of symbols

Sketching plant compositions

Technical planting plans do not communicate the lively nature of a planting bed. Plant symbols sometimes need to be graphically assembled to reflect variations and ideas for a planting scheme, and to give an impression of its overall visual effects.

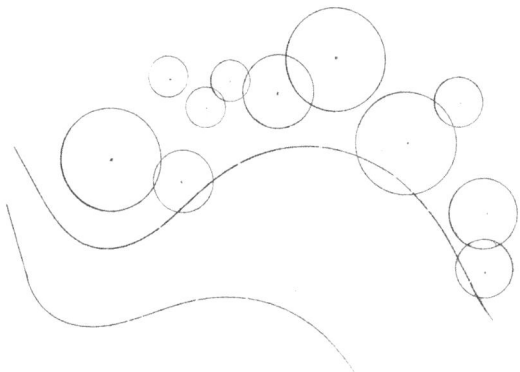

The pencil drawing assembles the important shrubs, leading perennials and ground cover areas. Plant sizes and placements can be quickly varied at this stage.

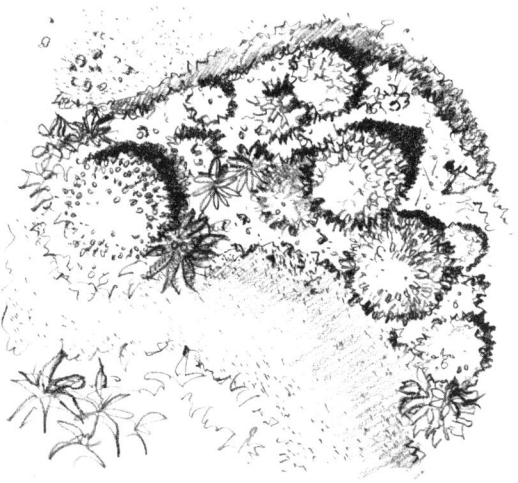

The sketch is further enhanced with symbols and textures. The closer they are to each other, the more dissimilar they need to be to ensure legibility. Foliage textures can be added to increase spatial volumes.

Adding shadows can give further graphic effect to the composition, allowing the differing heights of the individual plants to be understood

Please remember that the examples shown here are primarily concerned with the graphic impression of a planting design, and are not planting plans for actual construction

Trees and vegetation
Drawing trees
Tree symbols
Shade, shadows, and tonal values
Tree groups
Shrubs, hedges, and grass
Trimmed hedges and woody plants
Vegetation surfaces
Flowering plants
Sketching planting beds

3

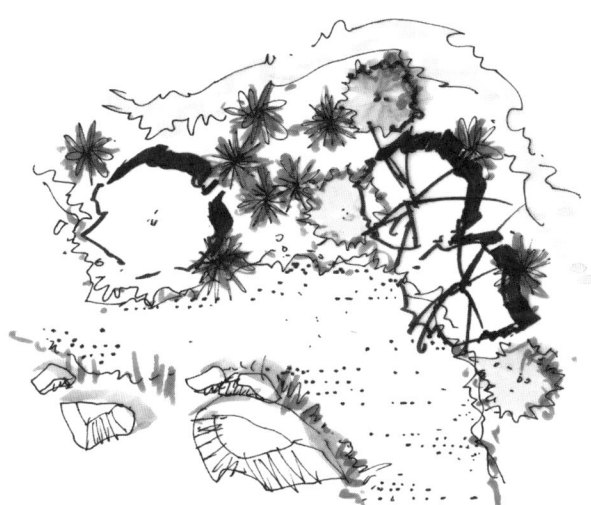

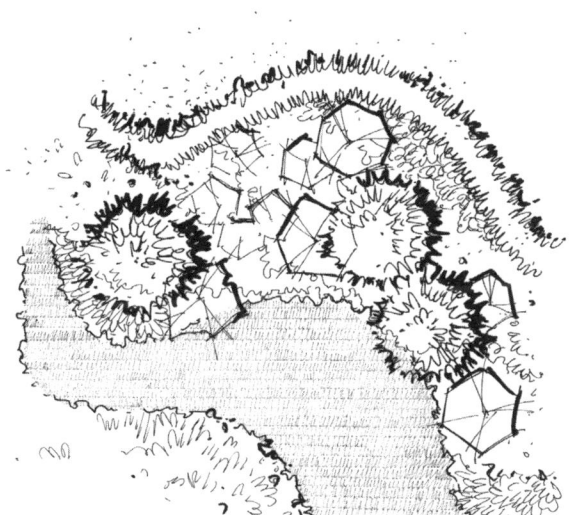

Contrasting shapes, line weights, textures and tonal values should all come together to form a visually balanced and easily legible whole

The faster the drawing, the less detail in the vegetation. The symbols, however, should still allow the viewer to distinguish one group of plants from another.

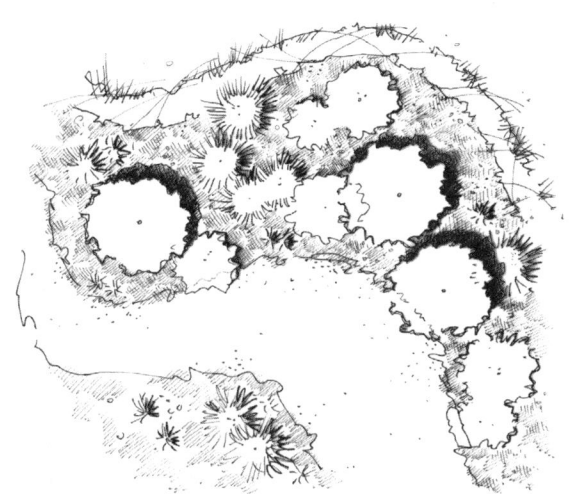

The plan view and the rendition of symbols

Drawing plants together

It is important to remember that simple lines and sketch techniques can only convey so much about plants. Vegetational structures are very complex and intricate. Unless we have lots of time to devote to artistic renderings of plants, it is usually sufficient to stick to the key features of the plants and to reduce them into legible, irregular textures.

In a sketch it's not so much the detailed realism that counts, as the overall impression and feeling of the drawing. On the next page are several very quick sketch exercises describing plants and planting beds. None of the vegetation is perfectly rendered, however the symbols and textures communicate just the right amount of key characteristics to indicate the plants.

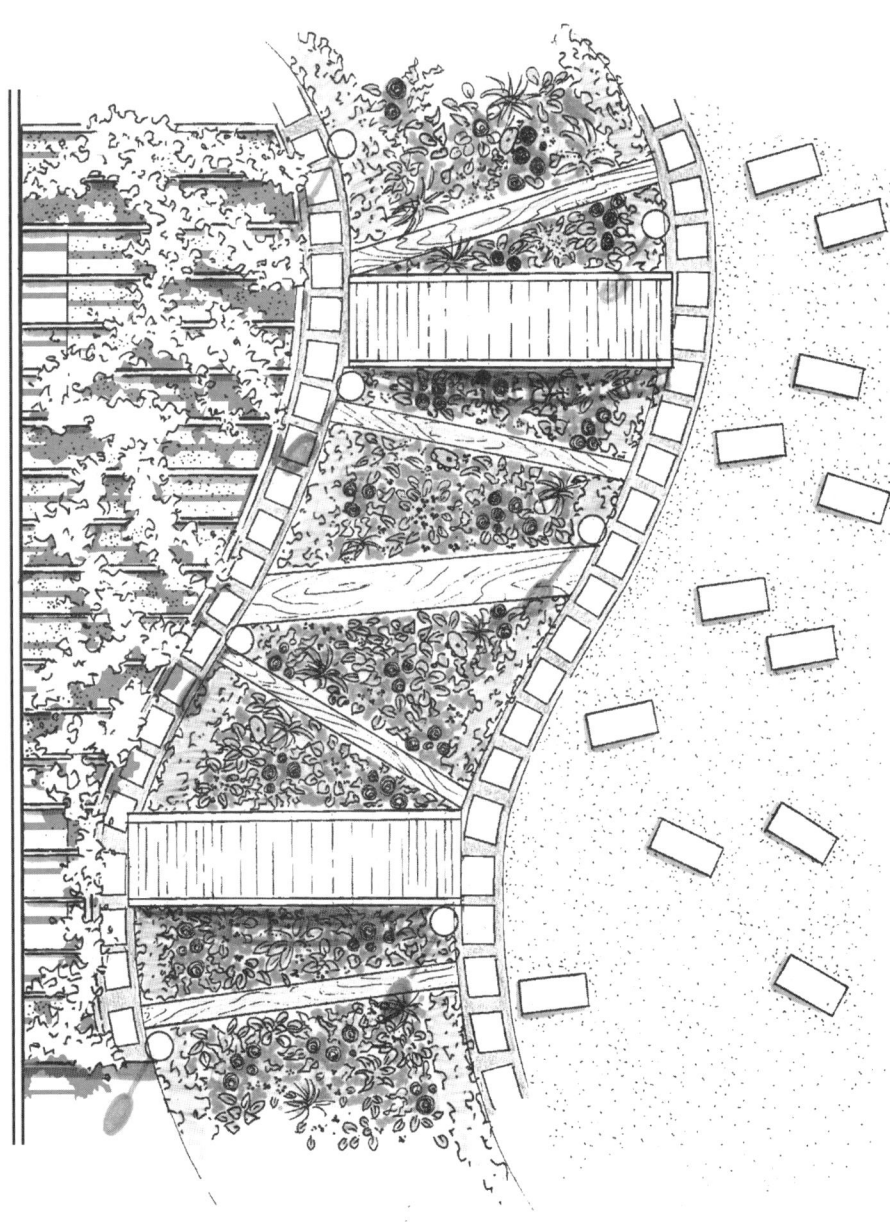

Trees and vegetation
Drawing trees
Tree symbols
Shade, shadows, and tonal values
Tree groups
Shrubs, hedges, and grass
Trimmed hedges and woody plants
Vegetation surfaces
Flowering plants
Sketching planting beds

3

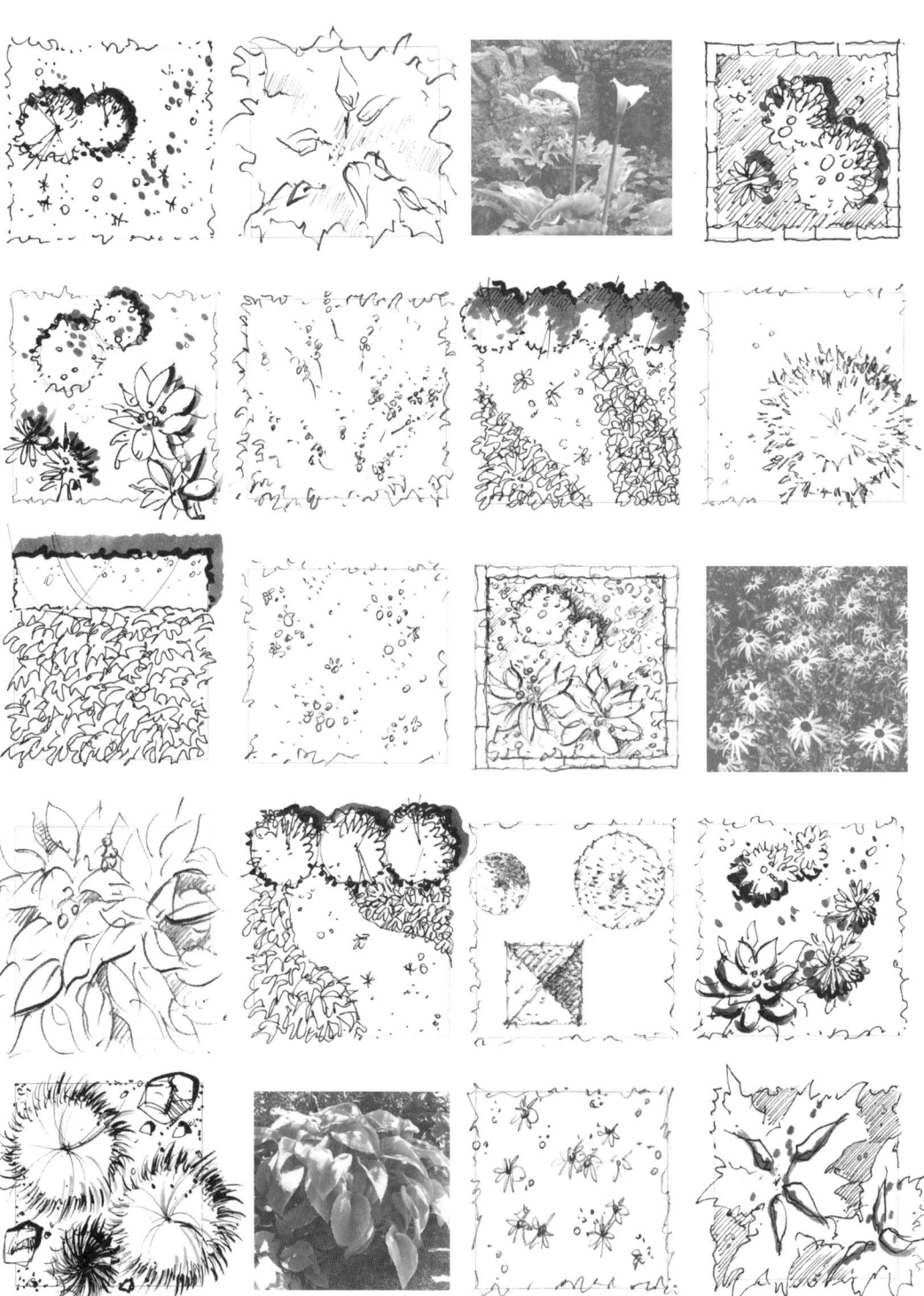

The plan view and the rendition of symbols

Drawing plants together

It can be daunting to draw vegetation. It is full of so many small, irregular and complex parts. A good way to practise is by closely looking at and recording plants from real life situations. Draw and sketch them as much as possible. Over time, sketching will automatically become faster and will concentrate on key features, leaving out much of the detail. Observation and freehand drawing remain the best ways to train perception and to gain confidence. They are also great ways to learn about plants and vegetation.

Trees and vegetation
Drawing trees
Tree symbols
Shade, shadows, and tonal values
Tree groups
Shrubs, hedges, and grass
Trimmed hedges and woody plants
Vegetation surfaces
Flowering plants
Sketching planting beds

3

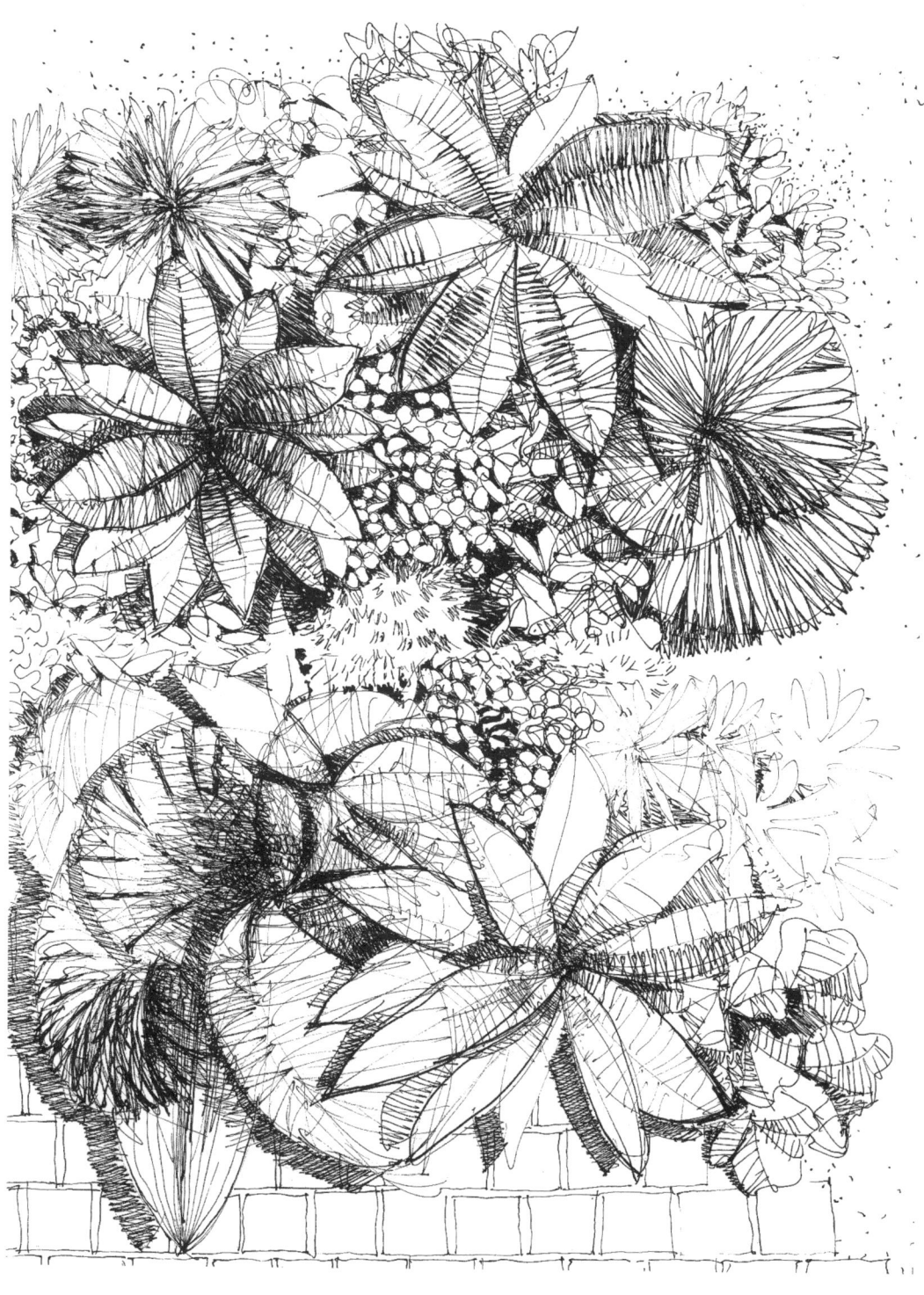

The plan view and the rendition of symbols

Pergolas, garden pavilions, and arbours

1:100 / 1:50

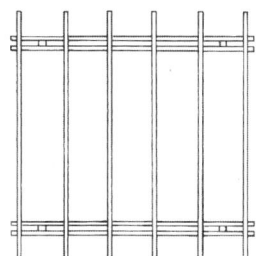

The individual parts of an arbour or pergola are represented by drawing their outlines

1:200 / 1:500

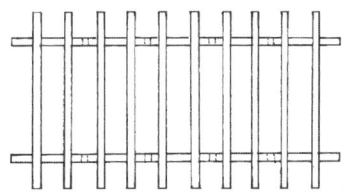

Depending upon the scale, the construction can be graphically simplified

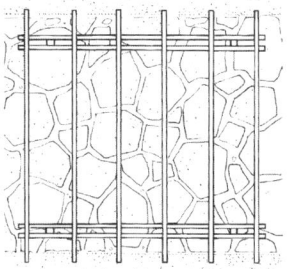

When elements overlap, certain elements may be blocked. Line weights are especially important to ensure that the pergola remains visible.

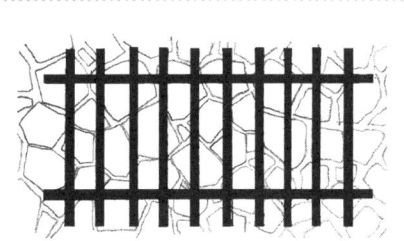

The pergola's construction elements should not be solid black. The drawing will appear flat and be difficult to read as a three-dimensional volume.

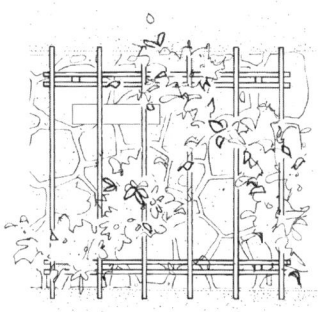

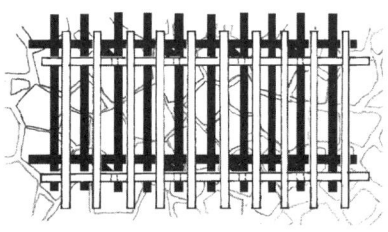

To increase the spatial effect, a slightly offset shadow will instantly lift the structure off the ground

Built structures
Pergolas, garden pavilions, and arbours

3

If the ground plane is deemed more important than the built structure, using dashed lines to indicate the contour of the overhead roof will communicate that it is present, despite the fact that it is not detailed

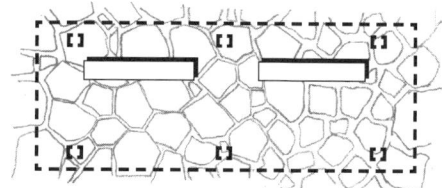

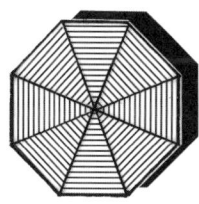 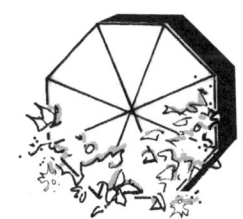  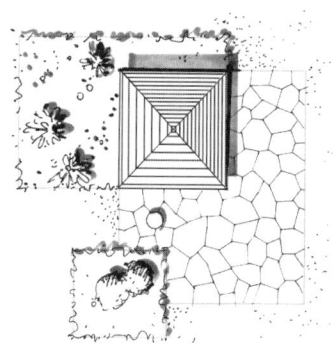

If a pergola or garden pavilion has climbing vegetation, this will block out some parts of the construction or roof plan

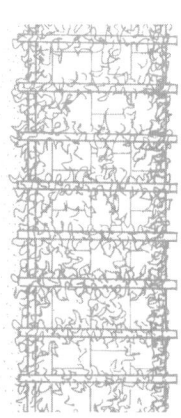

It is not necessary to cover every part of the pergola structure with vegetation. Uncontrolled scribbles all over the construction can result in a sloppy drawing, where neither the pergola nor the vegetation is clearly legible.

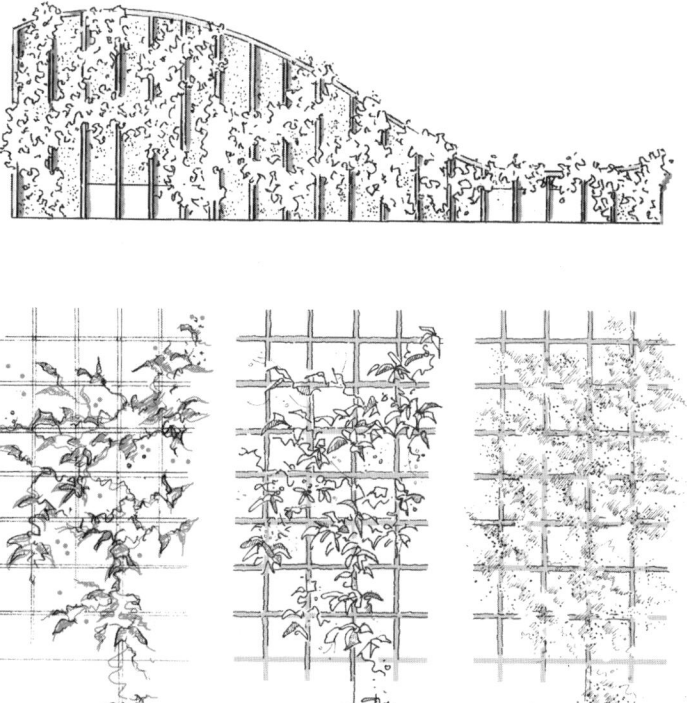

The plan view and the rendition of symbols

Pergolas, garden pavilions, and arbours
Pergolas, garden pavilions, arbours, and gazebos can have many different shapes. They are often combined with a seating area in a garden or park. Here are a few examples of such structures and shelters.

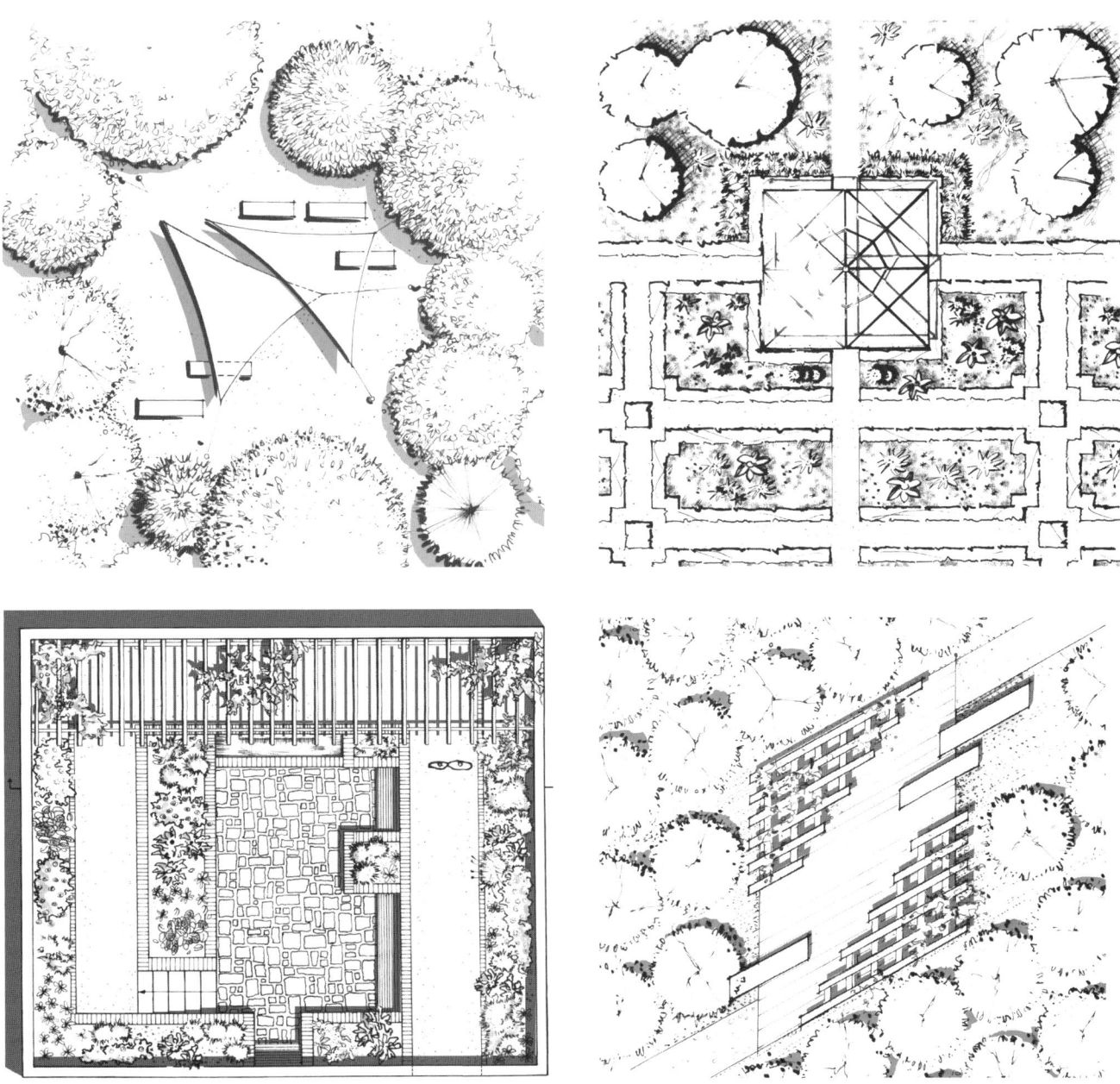

Built structures
Pergolas, garden pavilions, and arbours

3

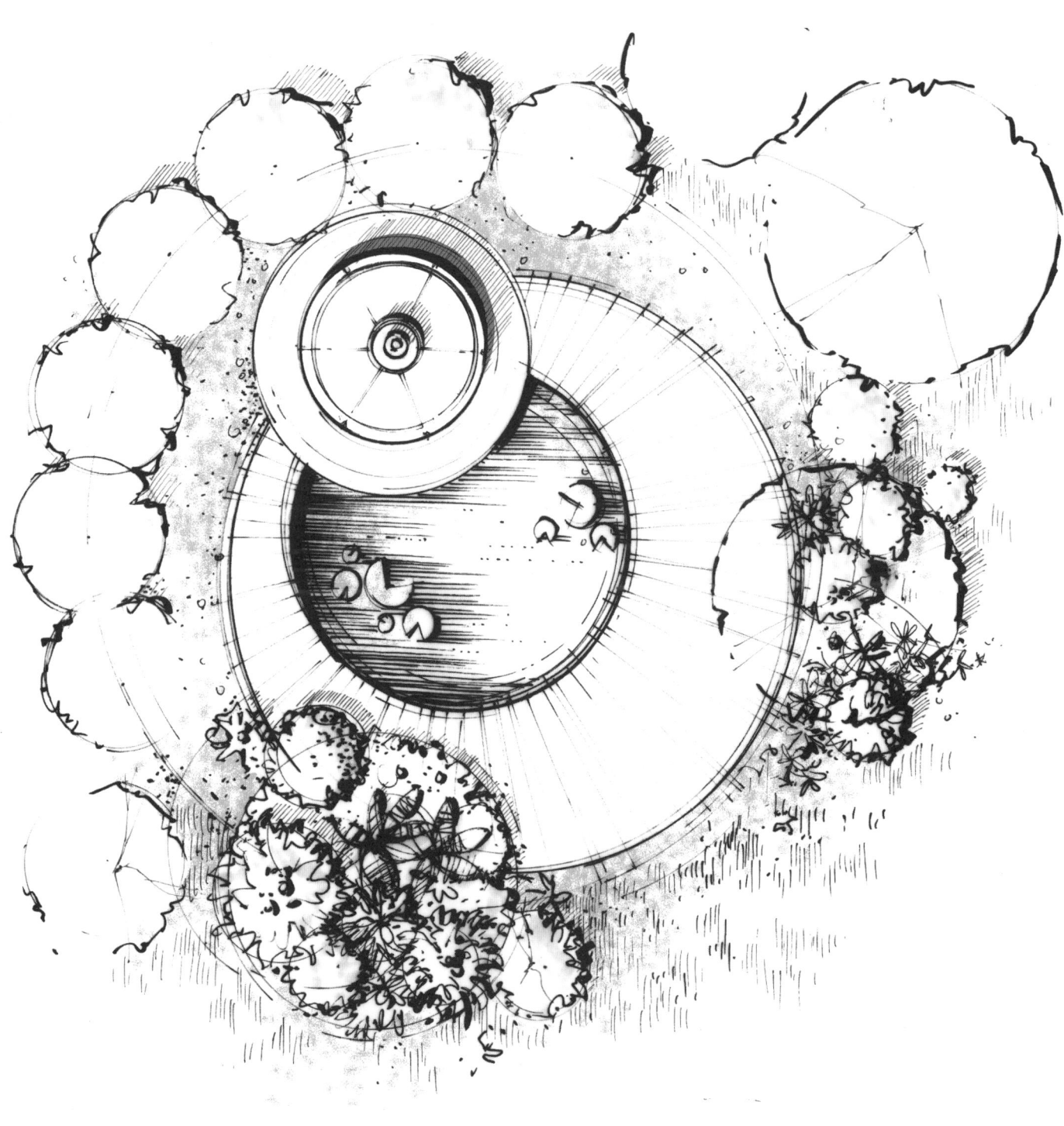

The plan view and the rendition of symbols

Drawing surfaces, paving patterns, and materials

1:50/
1:20/
1:10

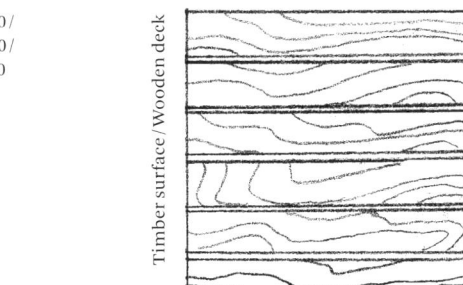 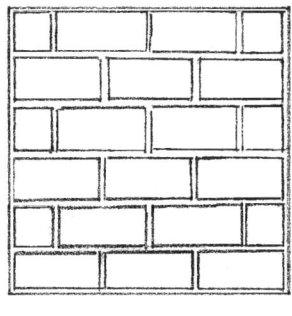 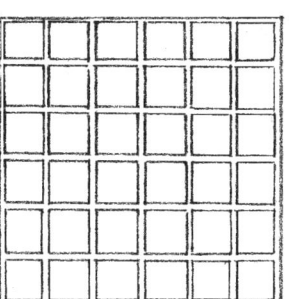

The more detailed the project, the more materials can be shown.
Different surfaces and their patterns become clearly visible and distinguishable.

1:100/
1:200

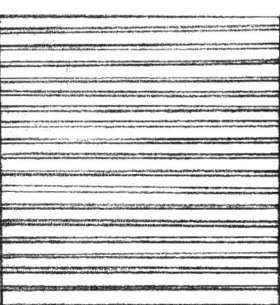 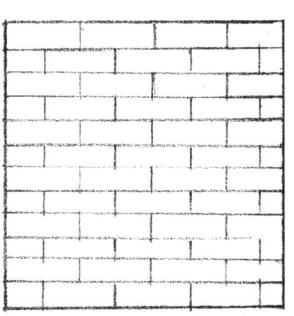 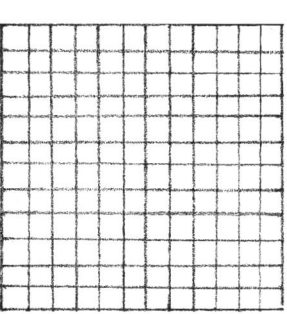

As the viewer moves further away from the ground plane, only overall patterns remain visible

1:100/
1:200/
1:500

 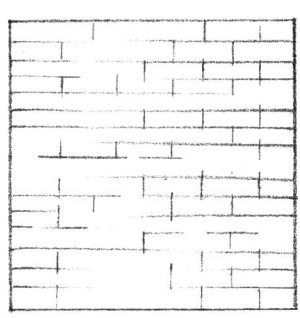 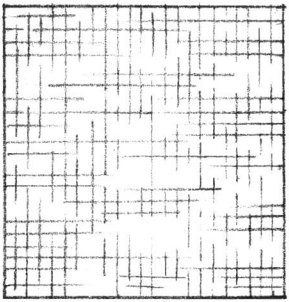

When zooming out, it is no longer necessary to draw and texture every part of the surfacing. Patterns and materials are hinted at around the edges of a plane, fading out towards inner areas.

Surfaces and materials
Paving patterns and scales
Freehand surfaces
Walls, stairs, and ramps
Rocks and stone walls

3

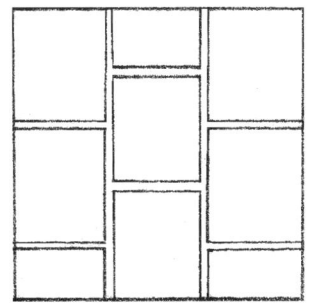
Rectangular pavers, regularly offset

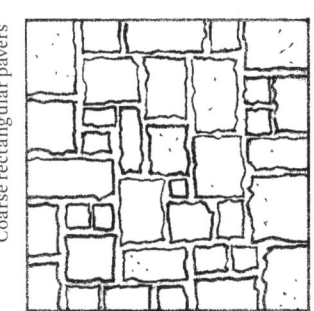
Coarse rectangular pavers

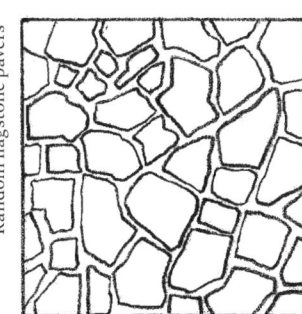
Random flagstone pavers

How much detail is drawn in the ground plane will depend on the project scale. As CAD programmes will provide very precise detailing, hand drawn surfaces need only to suggest the texture and paving pattern, leaving out or fragmenting the patterns towards the middle.

It takes a bit of practise to leave out areas of paving pattern and not also leave empty visual holes in the drawing. When in doubt, it's best to draw the entire paving pattern, remembering to use thin line weights.

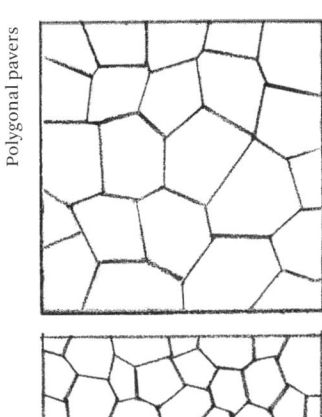
Polygonal pavers

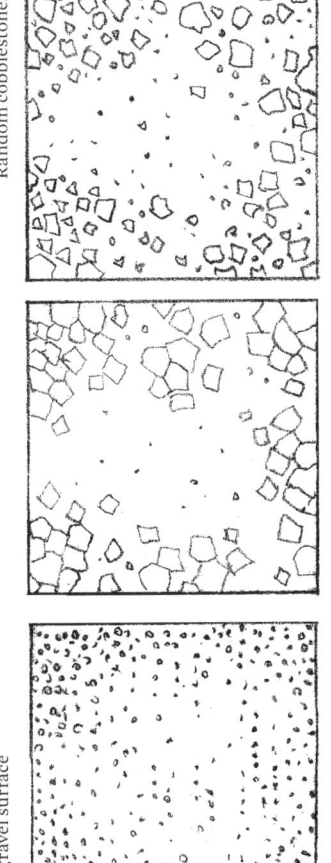
Random cobblestones

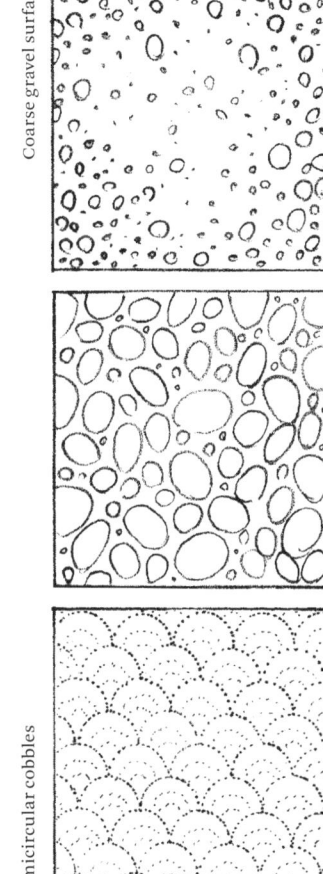
Coarse gravel surface

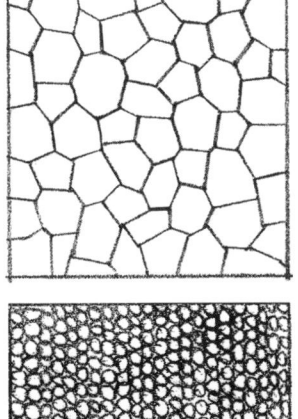
Cobblestones

Fine gravel surface

Semicircular cobbles

87

The plan view and the rendition of symbols

Drawing paved surfaces

Drawing the entire paving pattern or hard surface in a plan will take more time, but will require less thought. Any areas left without paving patterns or textures should be given carefully considered within the overall composition of the plan. The areas without detailing should not appear empty.

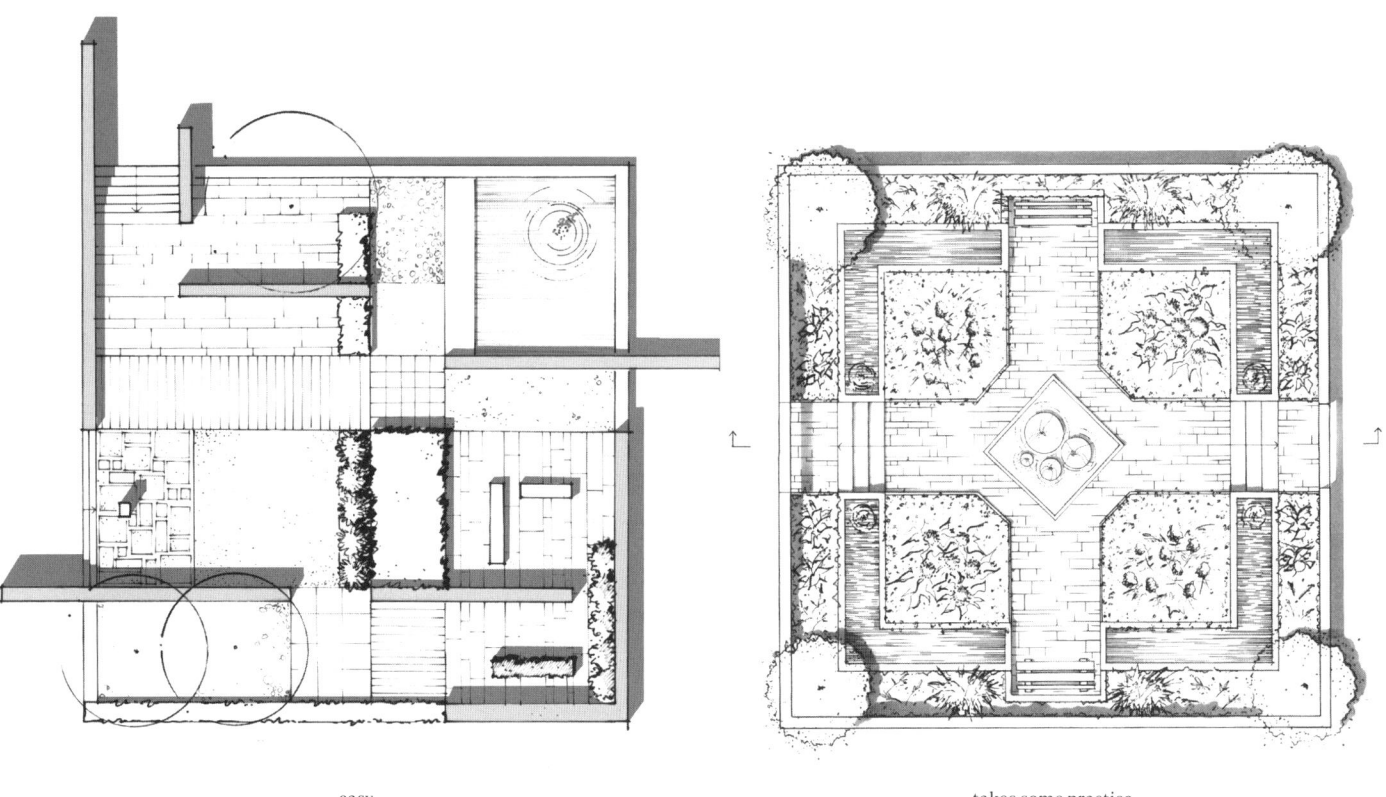

easy   takes some practise

Surfaces and materials
Paving patterns and scales
Freehand surfaces
Walls, stairs, and ramps
Rocks and stone walls

3

These two examples have their surfaces drawn in full. The different paving patterns and textures are clearly visible and are a part of the overall concept.

The smaller the paving stones, the darker and denser a surface area will appear. Surrounding elements must appear lighter in order to balance the overall visual effect of the composition.

The plan view and the rendition of symbols

Drawing walls, stairs, and ramps

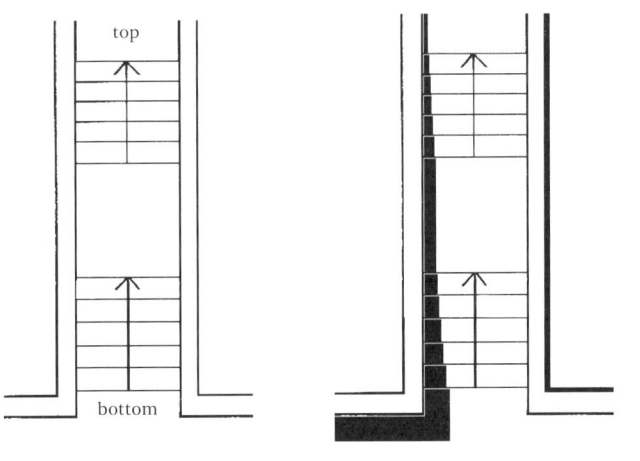

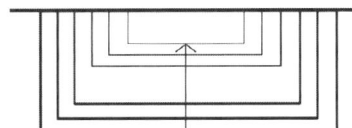

Stairs can take on many different forms. They are simple to draw and must include an arrow to indicate the top step. Adding shadows to the steps can also increase legibility.

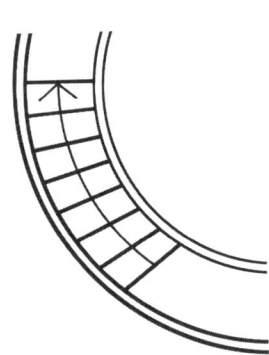

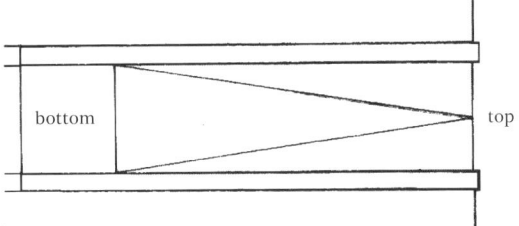

Ramps are also given an arrow to indicate the top of the sloped area.

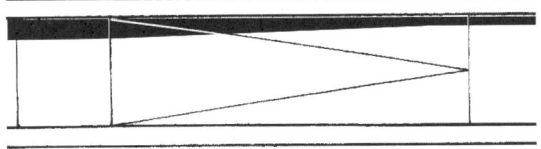

Do not infill walls with solid black. They will only appear flat and the area will not give any information about the surface materials.

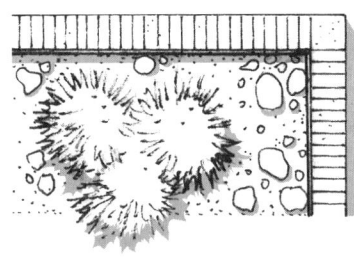

Surfaces and materials
Paving patterns and scales
Freehand surfaces
Walls, stairs, and ramps
Rocks and stone walls

3

Rocks and boulders

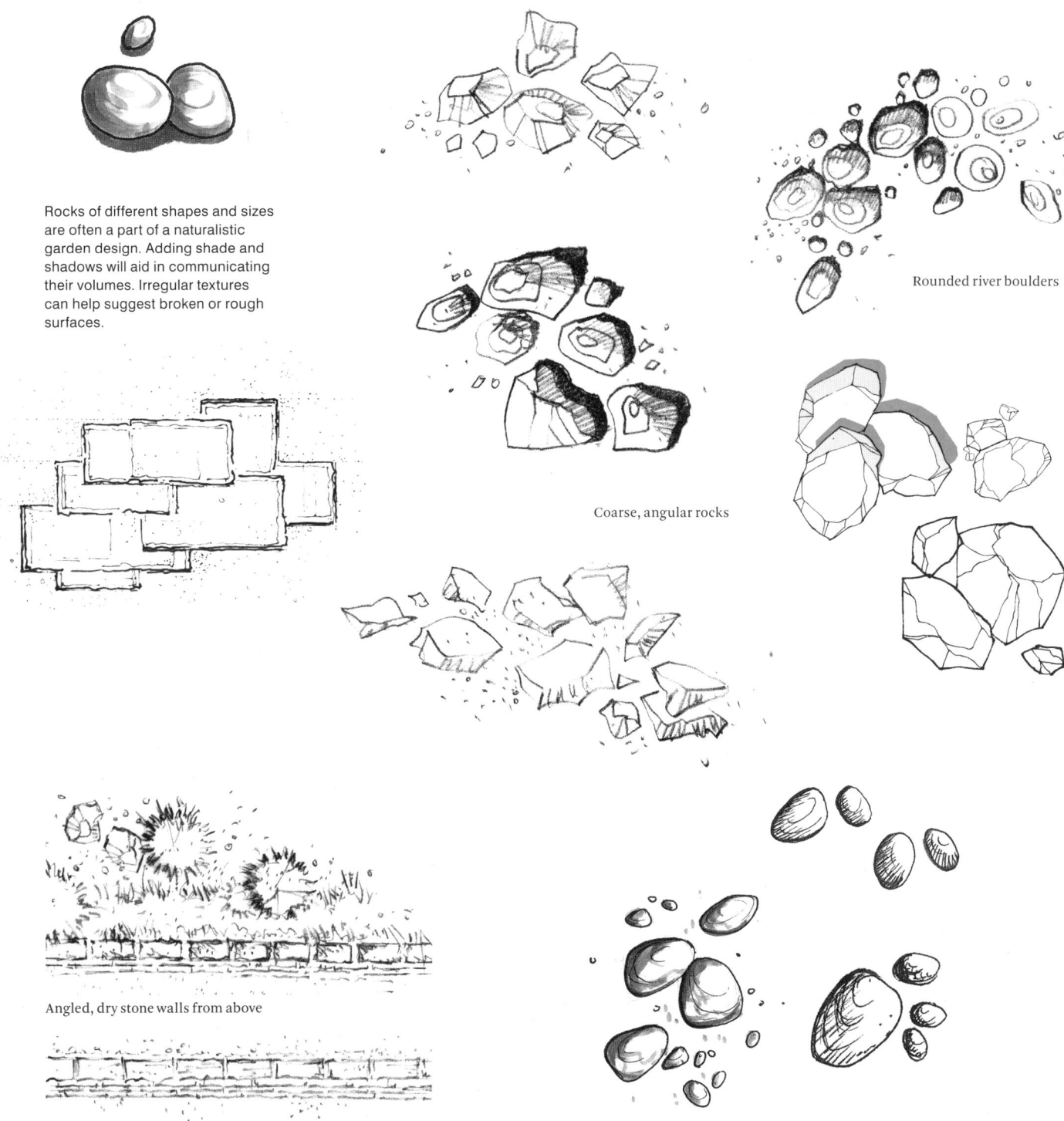

Rocks of different shapes and sizes are often a part of a naturalistic garden design. Adding shade and shadows will aid in communicating their volumes. Irregular textures can help suggest broken or rough surfaces.

Rounded river boulders

Coarse, angular rocks

Angled, dry stone walls from above

The plan view and the rendition of symbols

Built water features: Drawing water

Water is a particularly exciting and dynamic component of a garden or landscape design. It can take on many different forms: it can be calm and reflecting or flow vigorously. It can bubble, spray and fizz.

The most basic way of drawing water in a plan is by using fast horizontal strokes. These are applied irregularly, being denser around the edges of a water area, fading towards the middle. The quickest way to draw water is by using a straightedge or ruler.

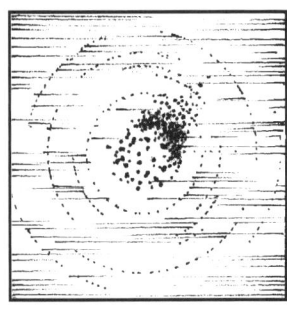 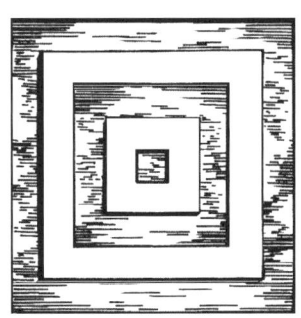 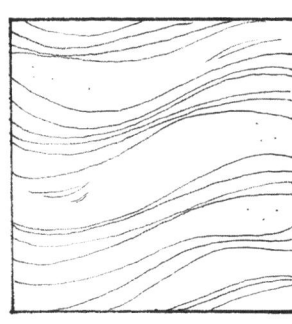 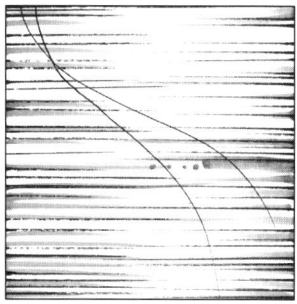

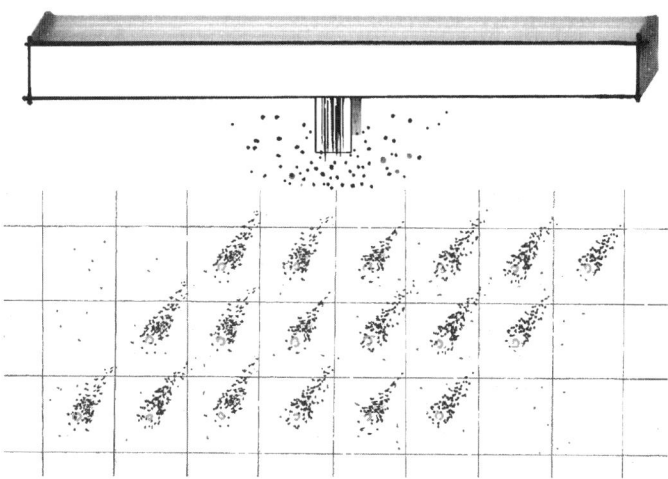

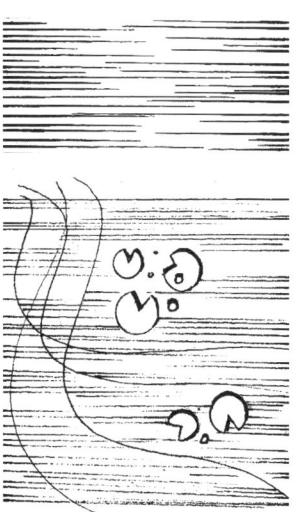

Moving water will create waves. These can be indicated with fine dotted lines within the water area. Water jets and spraying fountains can also be drawn using fine dots.

Water
Built water features
Moving water

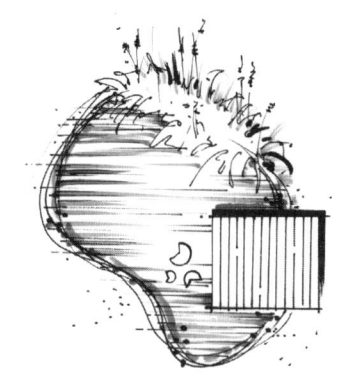
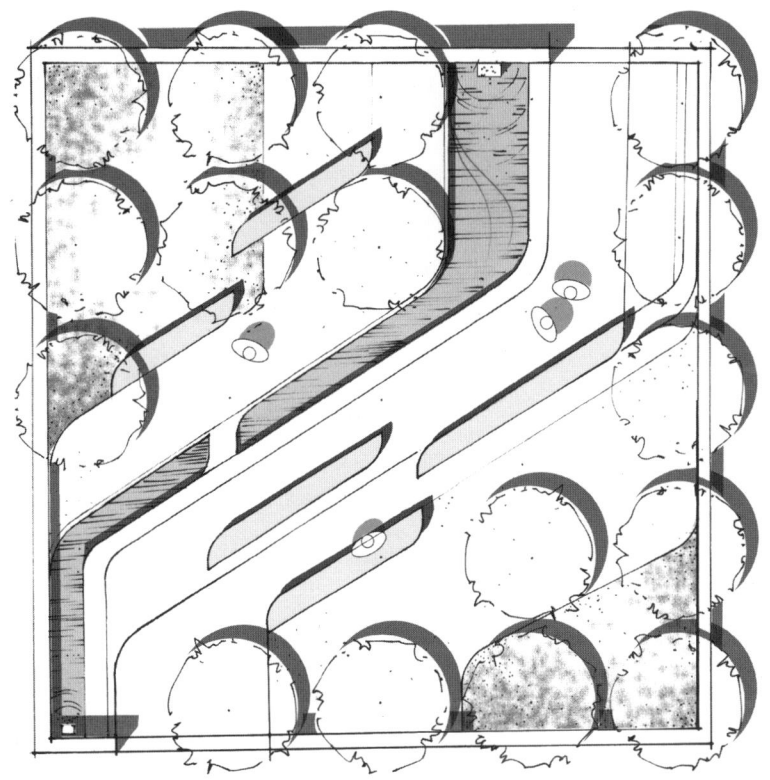
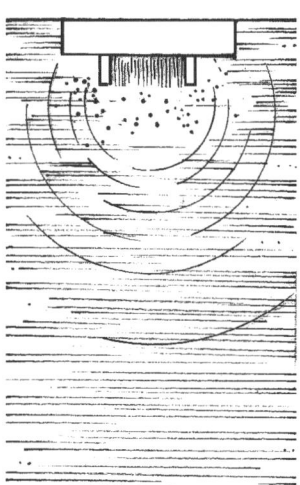
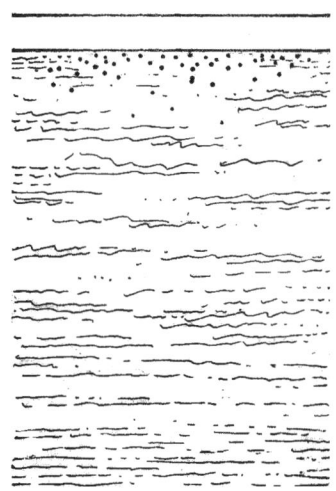

3

The plan view and the rendition of symbols

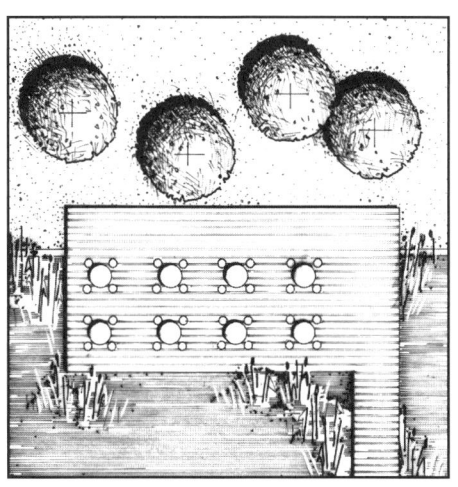

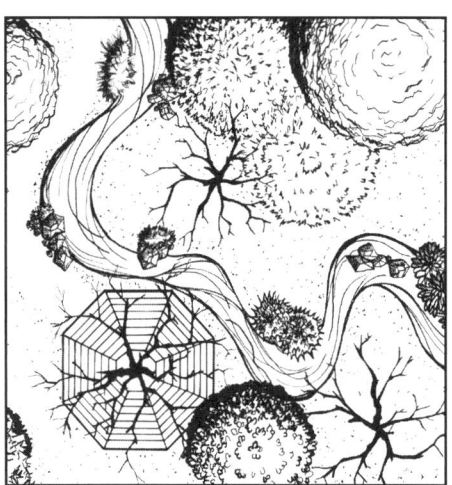

Water surfaces can be rendered as dark or light as needed. The tonal value of a water surface is a result of the density of the horizontal lines. The elements surrounding the water area usually determine how light or dark it needs to be rendered.

Water
Built water features
Moving water

Moving water

Flowing or meandering water features can be indicated simply by using curved lines. It is easiest to draw them freehand, however a flexible or curved ruler will achieve a neater effect. Overlapping these flowing lines within the water feature and along its edges can give a softer and somewhat more naturalistic impression.

Freehand flowing lines

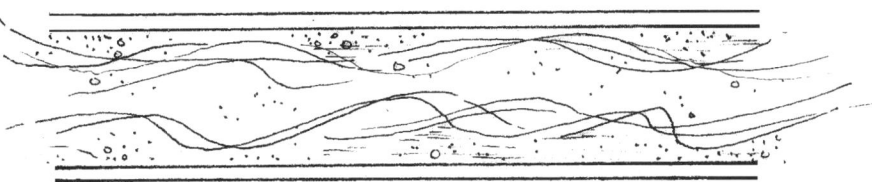

Flowing built water feature

Naturalistic water feature

The plan view and the rendition of symbols

Drawing furniture and people in plan

It is not always necessary to include people and furnishings in a plan. However, when communicating public spaces to others, indicating furniture and human figures can help to enliven a design and to show its functions and uses.

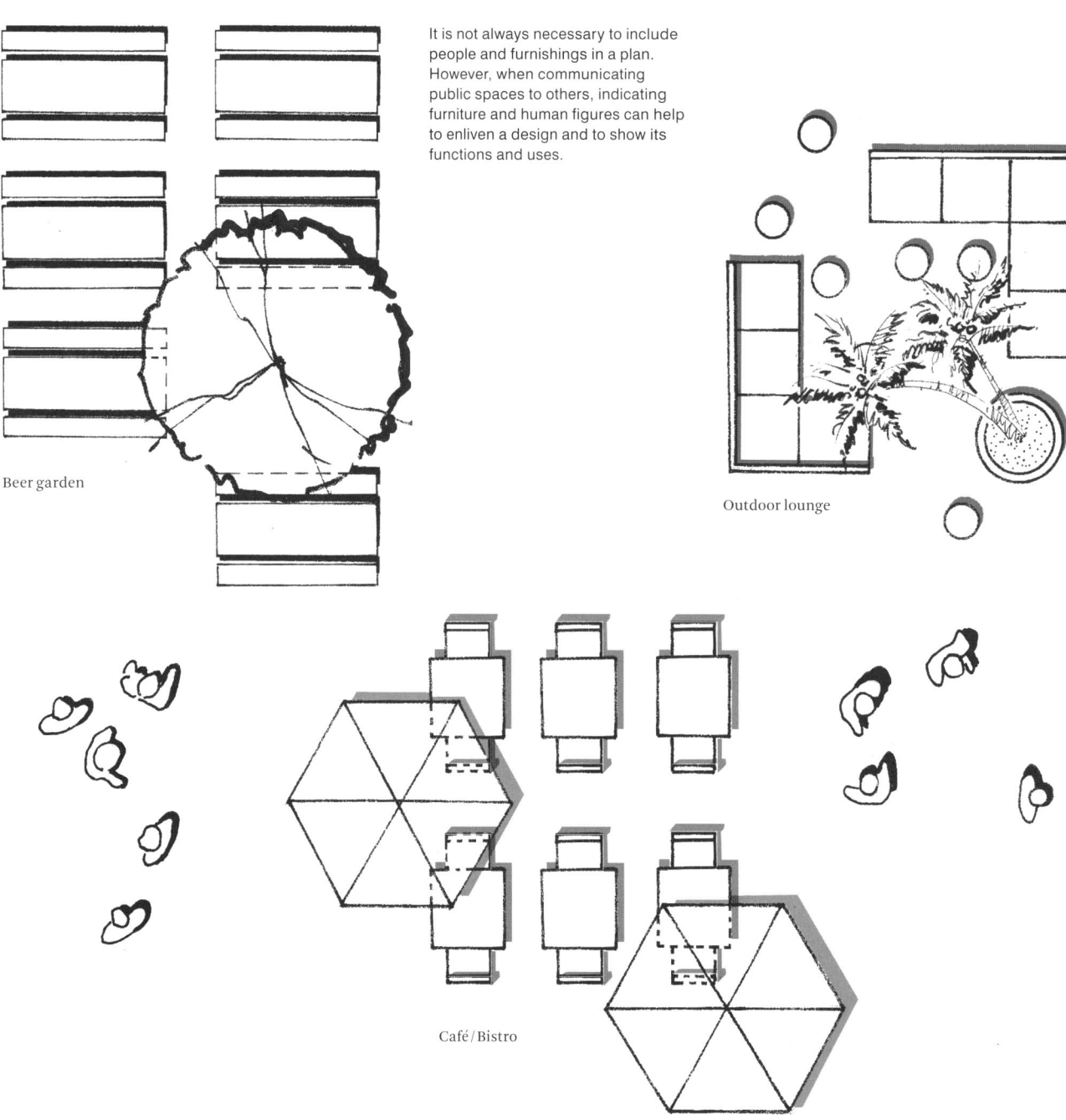

Beer garden

Outdoor lounge

Café/Bistro

Enliving scenes
Furniture and people

3

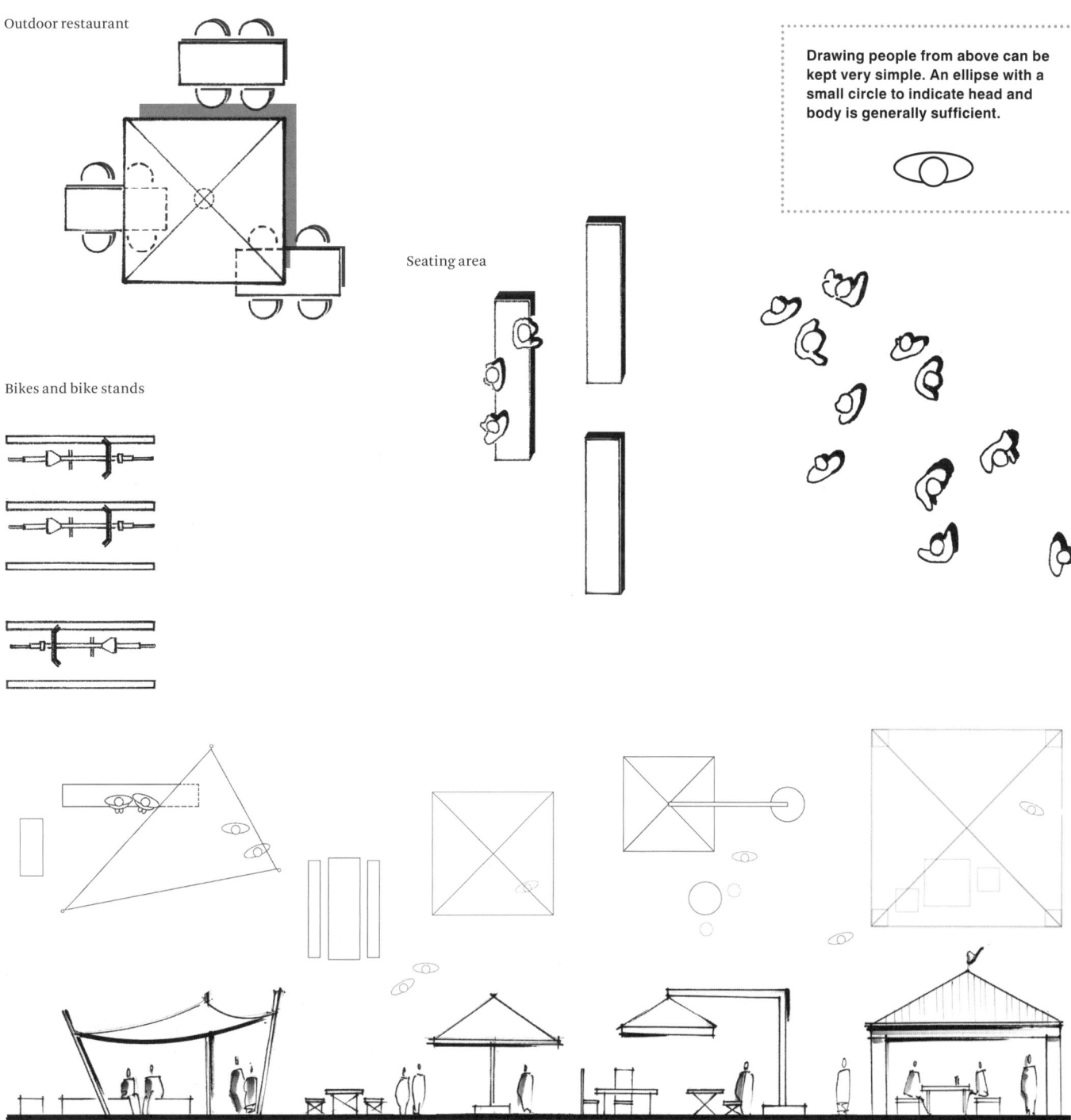

Outdoor restaurant

Bikes and bike stands

Seating area

Drawing people from above can be kept very simple. An ellipse with a small circle to indicate head and body is generally sufficient.

The plan view and the rendition of symbols

Topography, terrain, and landforms: Contour lines
When working on landscape projects, we often deal with natural landforms. These involve three-dimensional slopes and vertical changes, which have to be shown on a two-dimensional plan. It is extremely important to understand the concept of contour lines. These lines are a graphic method used to represent and communicate vertical changes and differing levels in topography to the viewer.

Each contour line represents a level elevation above or below a common measuring point. The lines will have defined intervals, usually determined by the topography or landform itself, as well as the scale of the drawing. For example, a very large site with steep slopes and embankments will have larger intervals than a smaller site with relatively flat terrain.

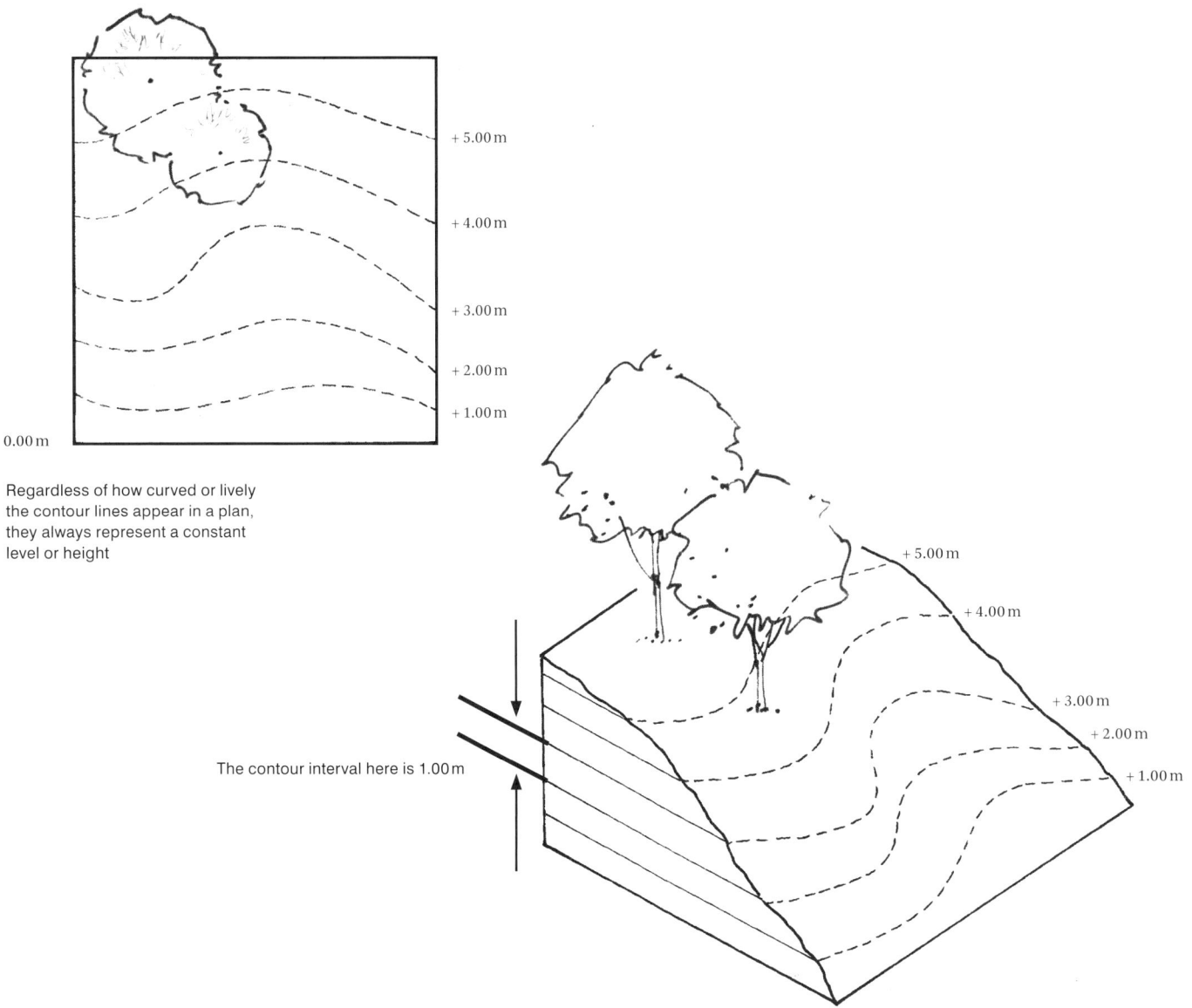

Regardless of how curved or lively the contour lines appear in a plan, they always represent a constant level or height

The contour interval here is 1.00 m

Topography and terrain
Contour lines
Retaining walls

3

These contour lines in plan represent a very uneven terrain

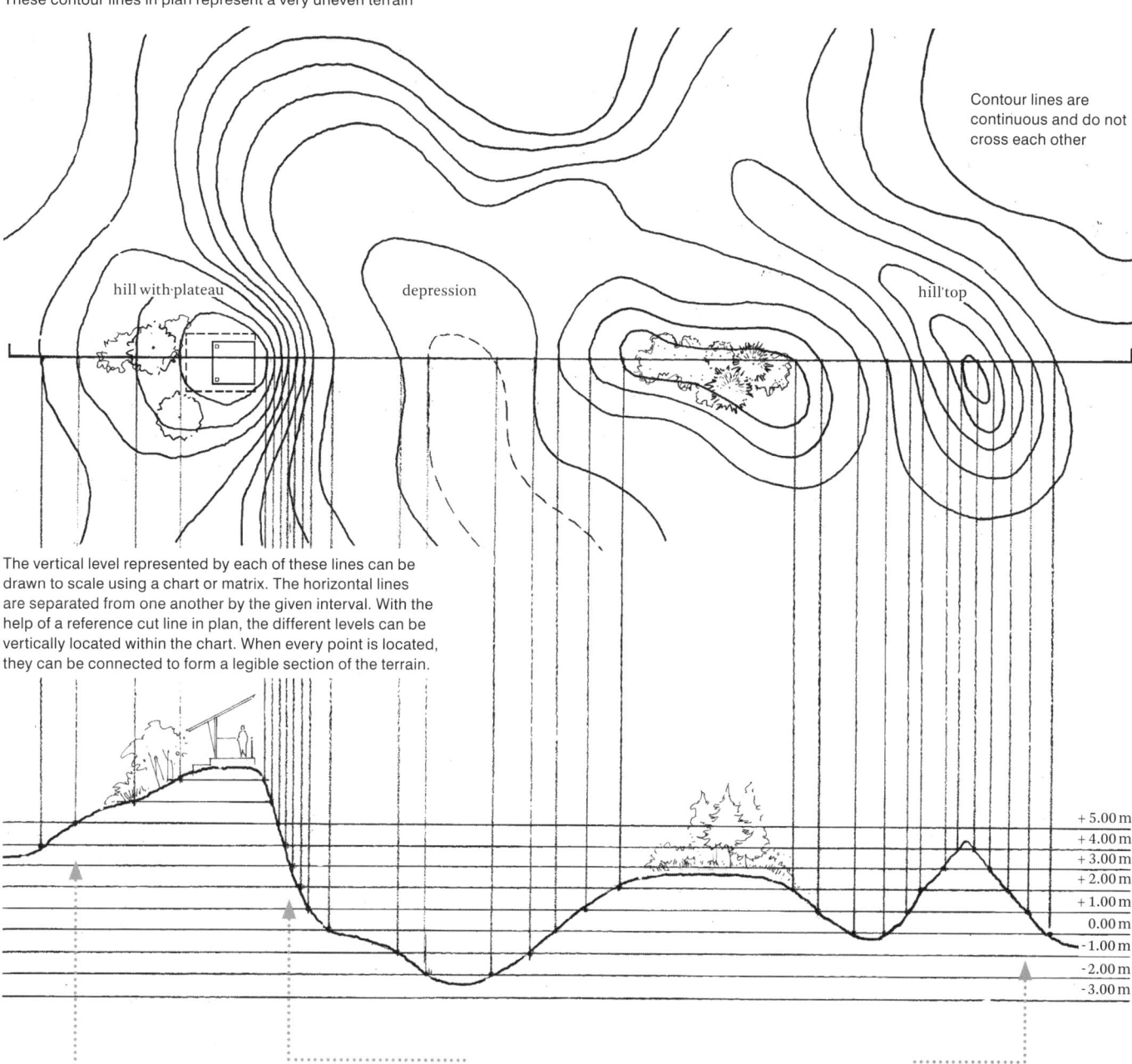

Contour lines are continuous and do not cross each other

hill with plateau   depression   hill top

The vertical level represented by each of these lines can be drawn to scale using a chart or matrix. The horizontal lines are separated from one another by the given interval. With the help of a reference cut line in plan, the different levels can be vertically located within the chart. When every point is located, they can be connected to form a legible section of the terrain.

+ 5.00 m
+ 4.00 m
+ 3.00 m
+ 2.00 m
+ 1.00 m
0.00 m
- 1.00 m
- 2.00 m
- 3.00 m

Widely spaced contours represent relative flat or softly sloped topography

Closely spaced contours represent steep slopes

Contour lines with relatively equal spacing will indicate a constant slope

See section on page 150

The plan view and the rendition of symbols

Contour lines should be drawn using thin line weights, as continuous, dotted or dashed lines. The contour interval the individual lines represent will be determined by the scale, size of site and the terrain itself. For smaller sites, the intervals can be 1 m apart or less. The steeper or larger the terrain, the larger the intervals will be. It is essential to indicate the intervals on or above drawn contour lines in plan, otherwise those lines will not be understood.

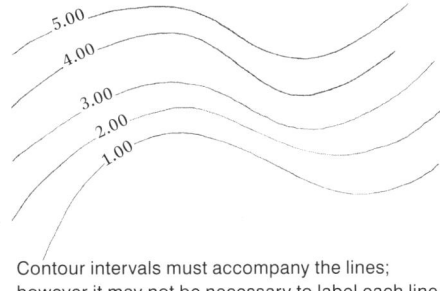

Contour intervals must accompany the lines; however it may not be necessary to label each line

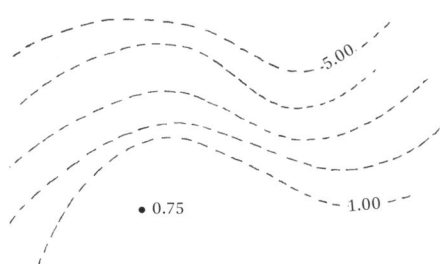

Spot elevations can lie outside of the contour lines and are indicated as necessary

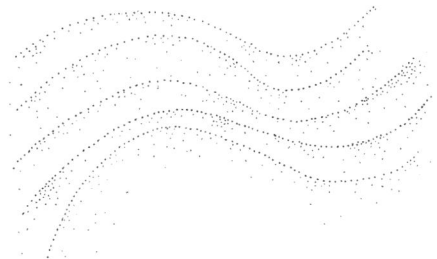

Dotted contour lines have a more subtle graphic effect than continuous lines

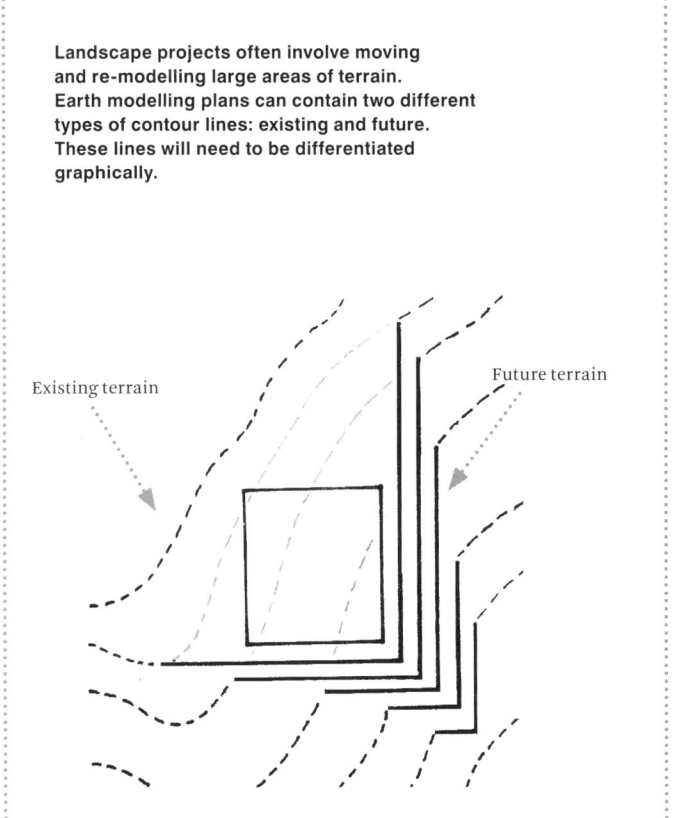

Landscape projects often involve moving and re-modelling large areas of terrain. Earth modelling plans can contain two different types of contour lines: existing and future. These lines will need to be differentiated graphically.

Topography and terrain
Contour lines
Retaining walls

3

Retaining walls

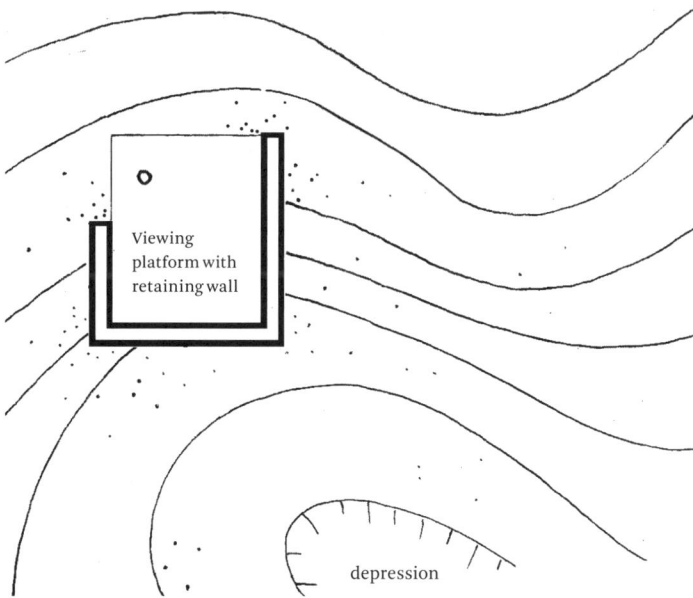

When contour lines meet up with a retaining wall, they seemingly disappear in plan.

In reality, the levels they represent simply continue across and along the vertical surface of the wall until they can continue at their level once again within the terrain.

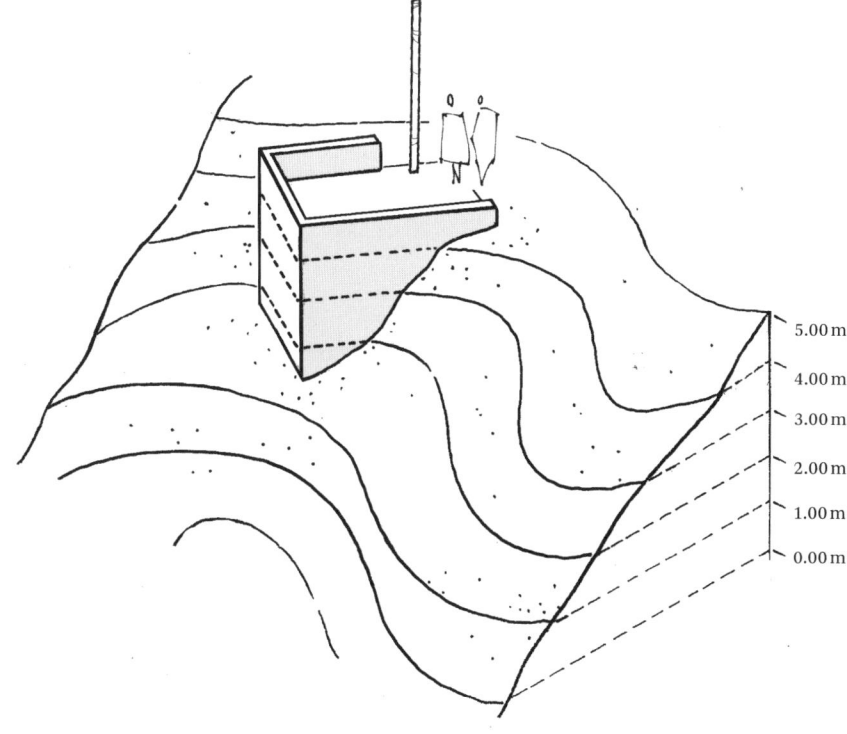

The plan view and the rendition of symbols

North arrows and graphic scales

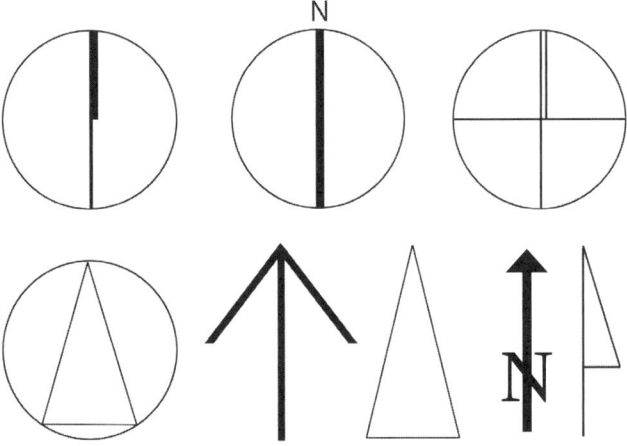

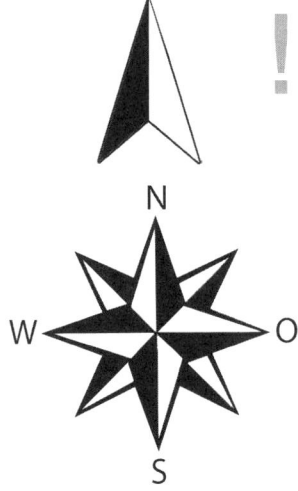

North arrows represent the orientation of the plan view. They are very important, especially if north is not directly vertical. North arrows generally face upwards; it is unusual for north to be facing down in a plan view.

Be careful with elaborate north arrows or very contrasting forms. They can quickly become too important and compete with other elements in the plan.

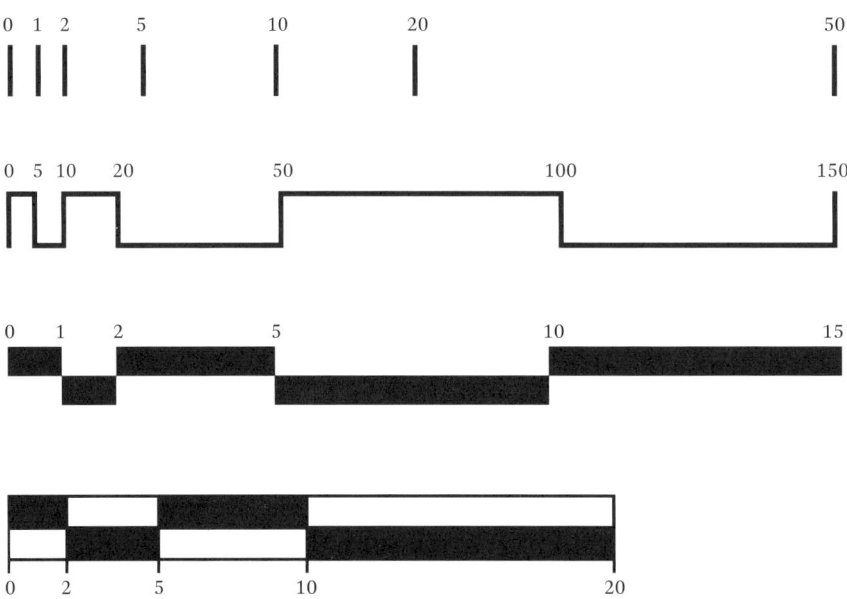

Graphic scales allow for a quick and easy estimate of dimensions in a plan view.

Especially when plans are not reduced or enlarged to scale, the graphic scales can communicate an easily understandable feel for lengths and proportions, as they will automatically be enlarged or reduced in proportion with the drawing. A written scale will not allow for this.

The intervals will depend on the scale of the project.

Graphic symbols
North arrow and graphic scales

3

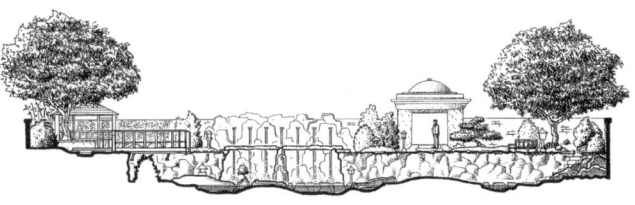

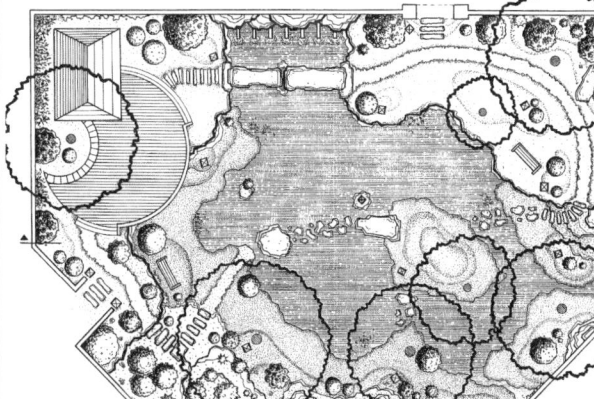

North arrows and graphic scales tend to lie outside the plan view. They should always be considered as part of the overall presentation and layout composition.

The plan view and the rendition of symbols

Putting everything together: The drawing process

The preliminary sketch provides the basis for trying out locations of elements, relationships, dimensions as well as graphic effects, textures and contrasts. When everything is thought through, the sketch gets redrawn using ink, usually onto vellum, which can later be reproduced onto paper. The drawing on this page uses foliage texture trees to highlight the semi-circular placement of the trees. The trees have a tonal value to give their unique semi-circular placement a stronger presence within the composition.

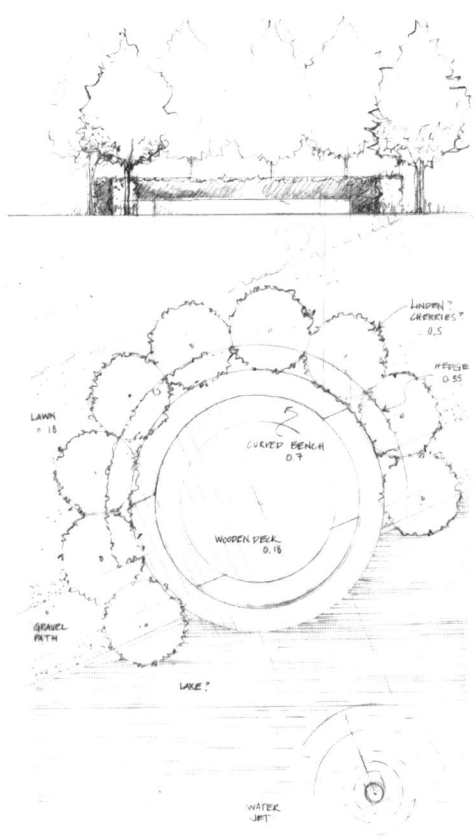

Drawings usually begin with a basic pencil sketch

Putting everything together
Drawing process
Elements of a successful line drawing

3

Since the planting beds and vegetation are an important part of the design scheme shown on this page, they need to be clearly visible in the plan. As a result, the large trees are kept transparent within the final drawing.

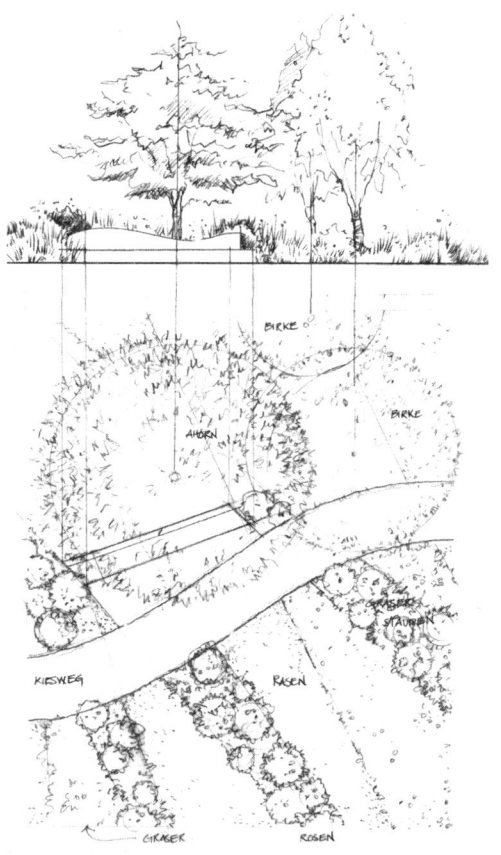

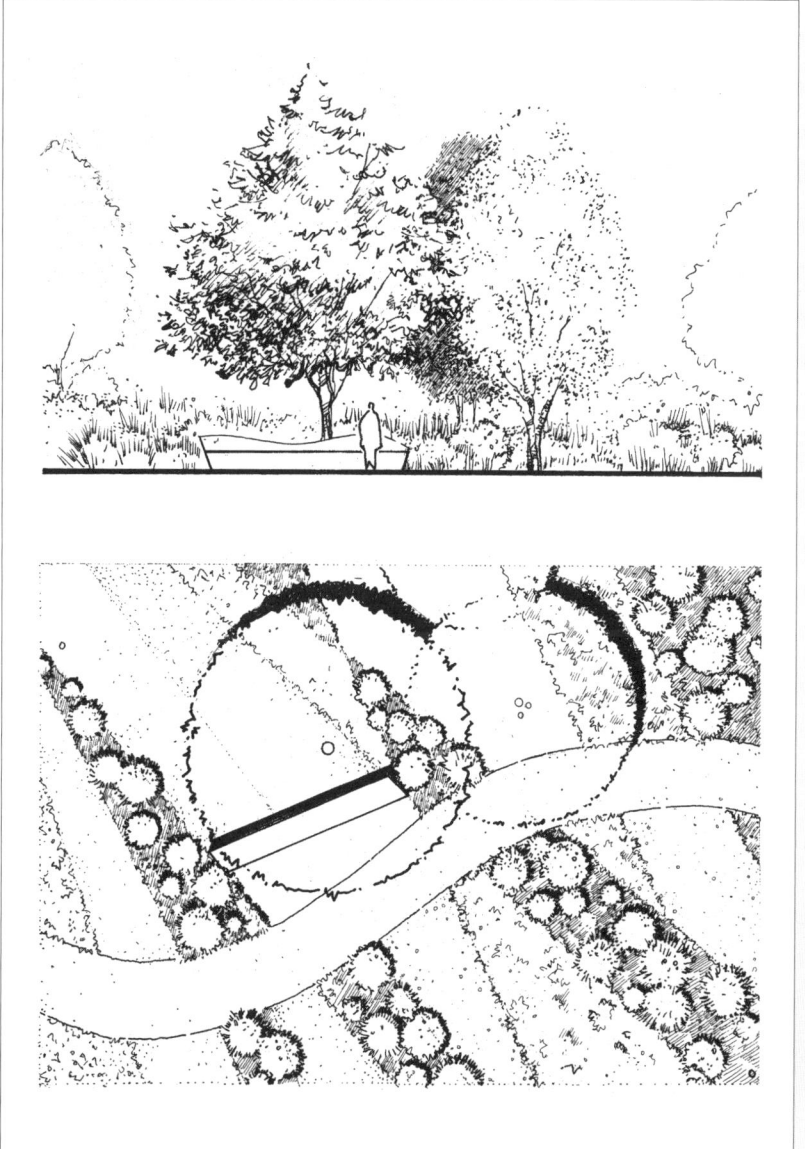

Even if trees are shown as transparent, they will still need a strong line weight for their outline. A shadow will lift the tree from the ground plane and prevent it from becoming invisible within the composition.

105

The plan view and the rendition of symbols

Elements of a successful line drawing
Lines form the basis of architectural drawings and graphics. When putting lines together in finished drawings, it is important to order and organise information so that it all works together in a successful way. Although there is no single foolproof way to ensure legibility and graphic effect, it is important to keep in mind the following points:

• Forms and outlines
Forms, shapes and their graphic outlines help to define and communicate objects, areas and surfaces in a drawing. They need to correspond to the object itself, allowing the viewer to quickly read and identify what is intended. Vegetation is thus drawn with irregular and small-textured outlines as opposed to hard, linear and orthogonal forms. These are best reserved for built structures and surfaces.

• Line weights and hierarchies
Spatially relevant objects need to jump out at the viewer within a drawing. This is easily done using stronger light weights for higher, more dominant elements. Trees and built structures should generally be drawn with a stronger line weight than the ground plane. Exceptions are possible, especially when the emphasis of the plan is the ground plane itself.

• Contrasts in surfaces and textures
Along with the line weights, contrasts in textures and their tonal values help with the legibility of different objects and areas in a plan. If two adjacent surfaces do not contrast enough, the viewer may not be able to distinguish them.

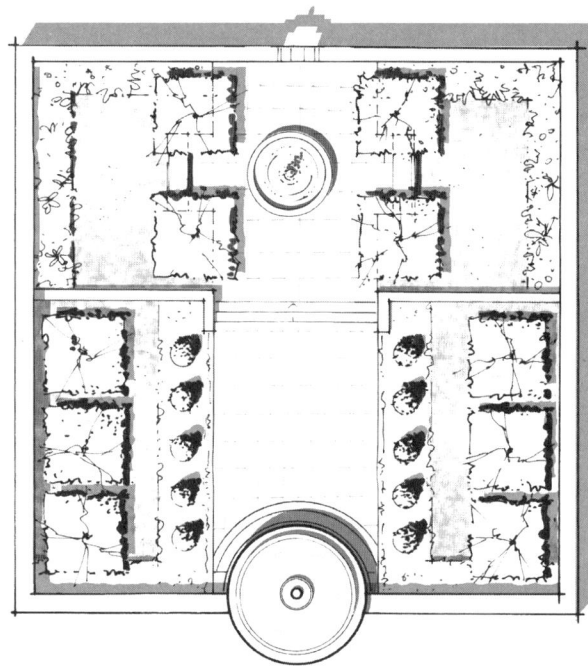

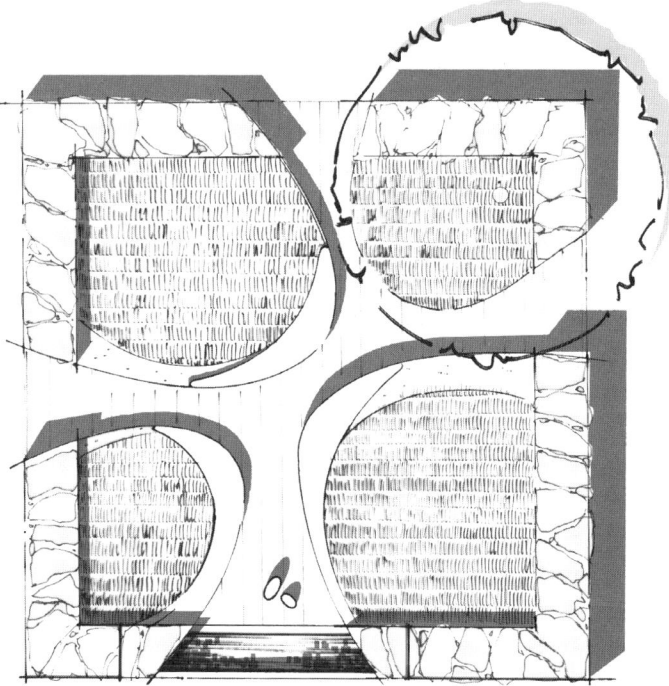

Putting everything together
Drawing process
Elements of a successful line drawing

3

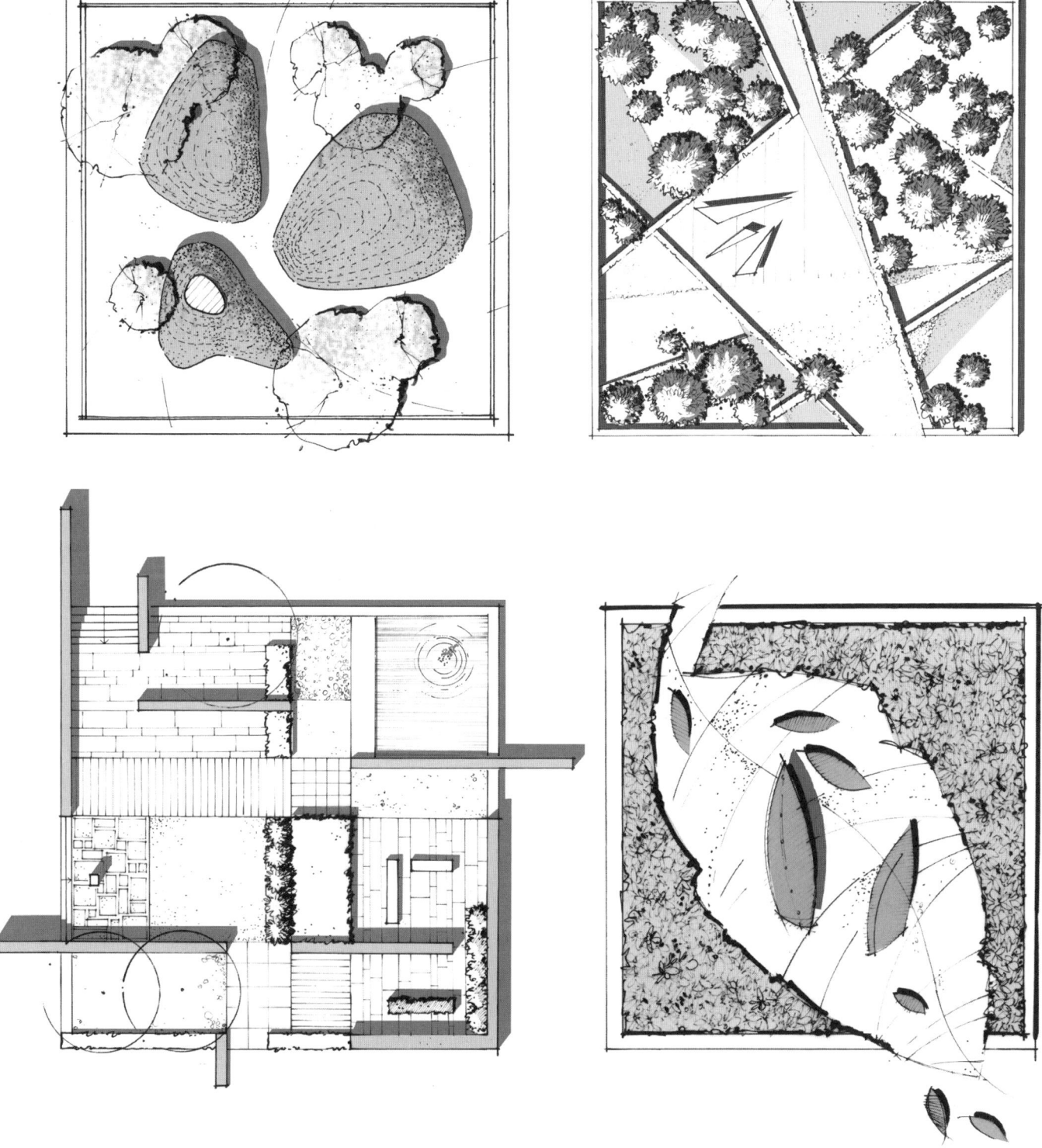

Elevation and section

Elevation
110 **Introduction**
112 **Construction**
Vegetation in elevation
116 **Trees**
128 **Shrubs and woody plants**
129 **Potted plants**
130 **Ground cover, grasses, and flowering plants**
132 **Adding depth**
Built structures
134 **Pergolas, pavilions, and arbours**
136 **Walls and materials**
138 **Water**
139 **People**
Section and section-elevation
140 **Introduction**
142 **Sections through buildings**
144 **Section cut lines**
145 **Constructing a section**
146 **Uses and scales**
148 **Section cut area**
150 **Examples**

Elevation and section

Elevation

Constructing an elevation involves projecting all visible plan content onto an upright picture plane (*see diagram on page 33*). The viewer looks at the scene before him, from a fixed position in the plan. Like the plan views, elevations are abstracted projections. Elevations are commonly found in architecture, where they are useful for communicating information about building facades and vertical elements, which may not be prevalent in a plan. An elevation begins with a principal sightline, from which the objects in plan are projected. It is best to place the sightline parallel to a façade in order to minimise distortion. Everything behind the line is then projected forward onto the picture plane. It is important to construct an elevation so that the most important design contents are included in the final elevation. A second parallel baseline can be drawn above or below the plan. This is, or is part of, a ground plane upon which the elevation is constructed. With the help of guidelines, the outlines of buildings and forms seen in the plan are projected onto the baseline. The heights and dimensions can then be drawn to scale (for example, the height of a roof or a tree). These points are then connected to form the elevation.

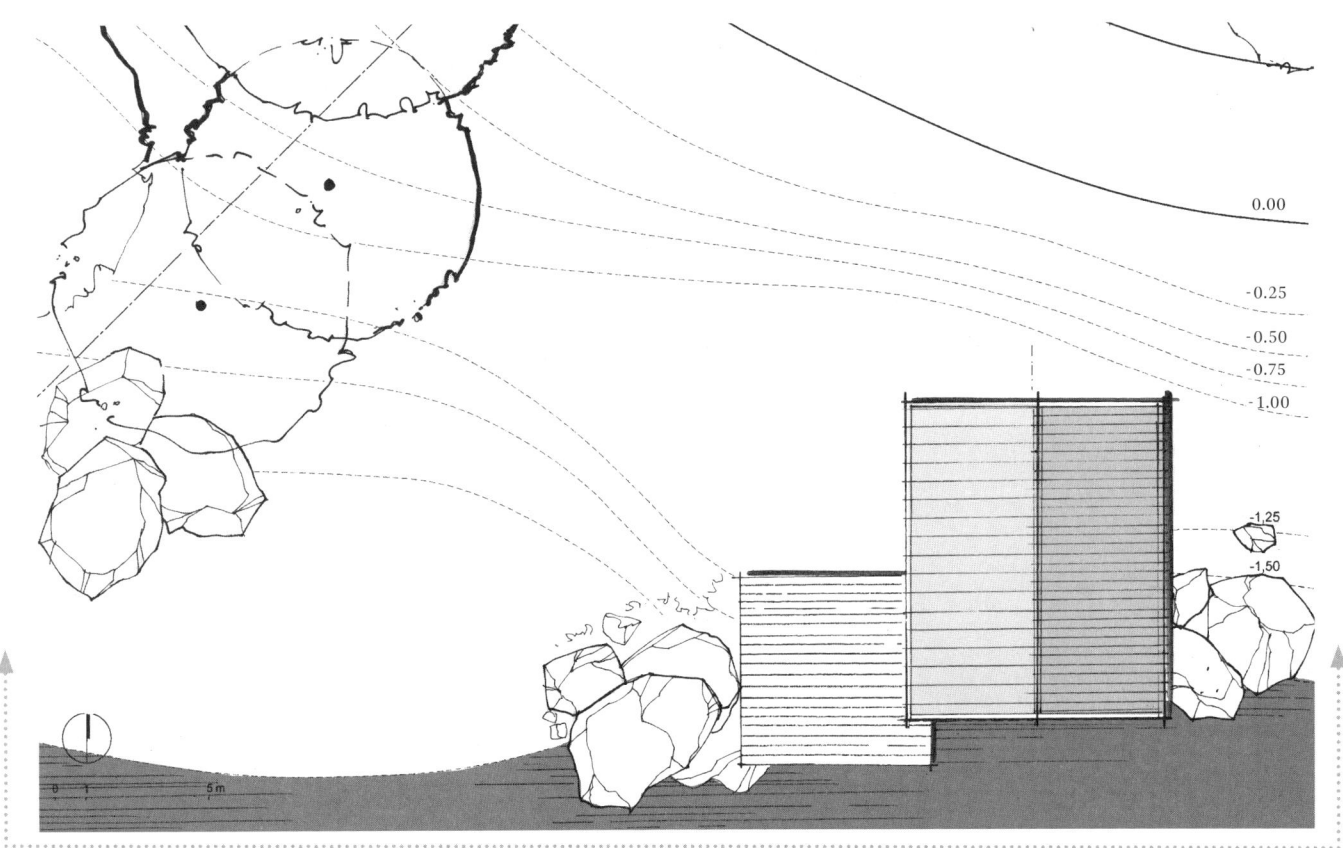

Elevation sightline which defines where the projected scene begins

Elevation
Introduction
Construction

4

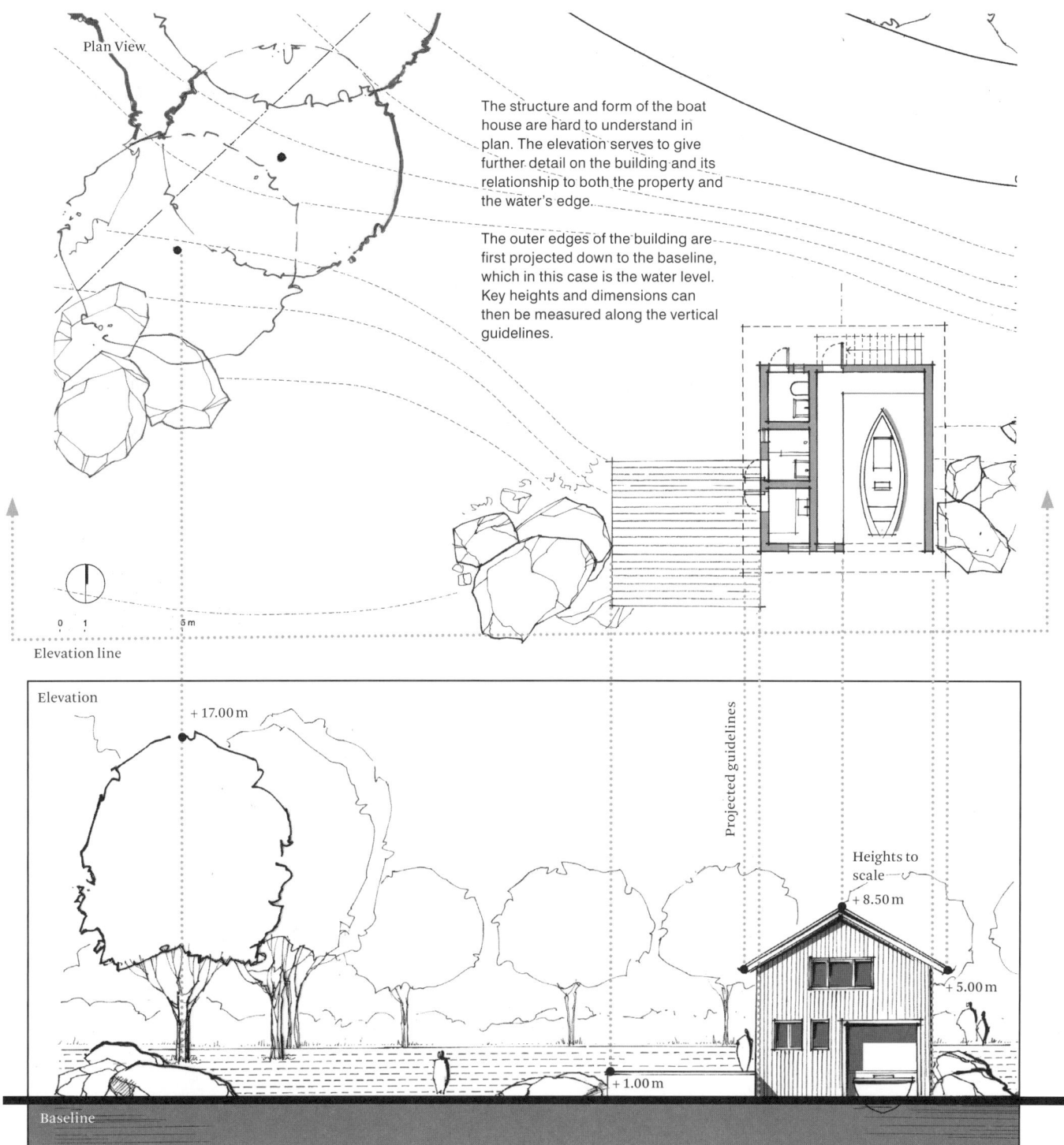

Plan View

The structure and form of the boat house are hard to understand in plan. The elevation serves to give further detail on the building and its relationship to both the property and the water's edge.

The outer edges of the building are first projected down to the baseline, which in this case is the water level. Key heights and dimensions can then be measured along the vertical guidelines.

Elevation line

Elevation

+17.00 m

Projected guidelines

Heights to scale
+8.50 m

+5.00 m

+1.00 m

Baseline

Elevation and section

Constructing an elevation
The point from which an elevation is to be drawn depends on what needs to be visually communicated. Sometimes it may not be a good idea to simply project the entire plan. The result may appear cluttered and difficult to read. Start the elevation at a key point, omitting what may not be relevant.

It may become necessary to draw more than one elevation, projecting in different directions, as was done in the examples on this page. The elevation on the next page begins down at the lake area and includes everything in the garden. The resulting scene is full and probably less legible to an untrained viewer.

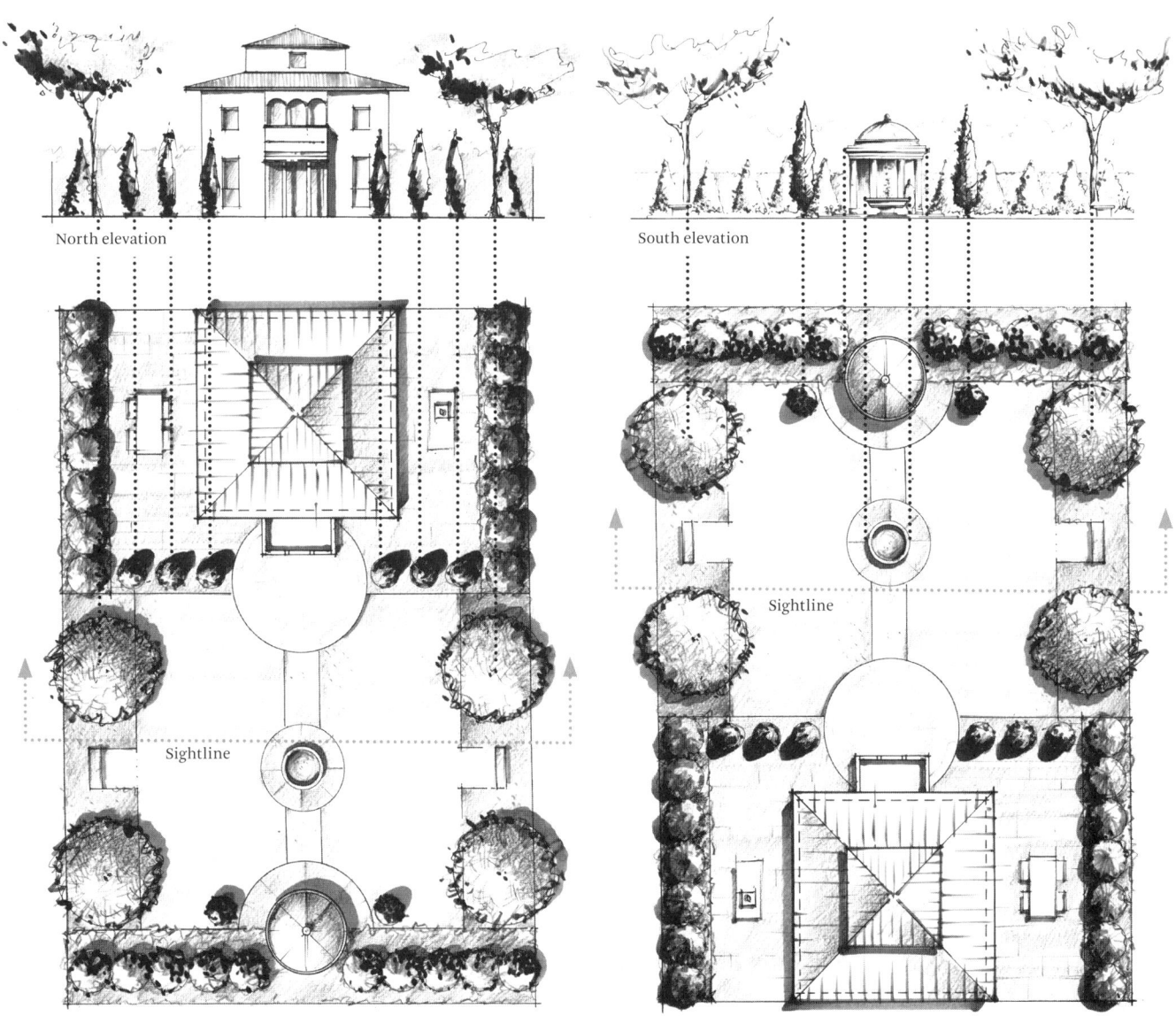

Elevation
Introduction
Construction

4

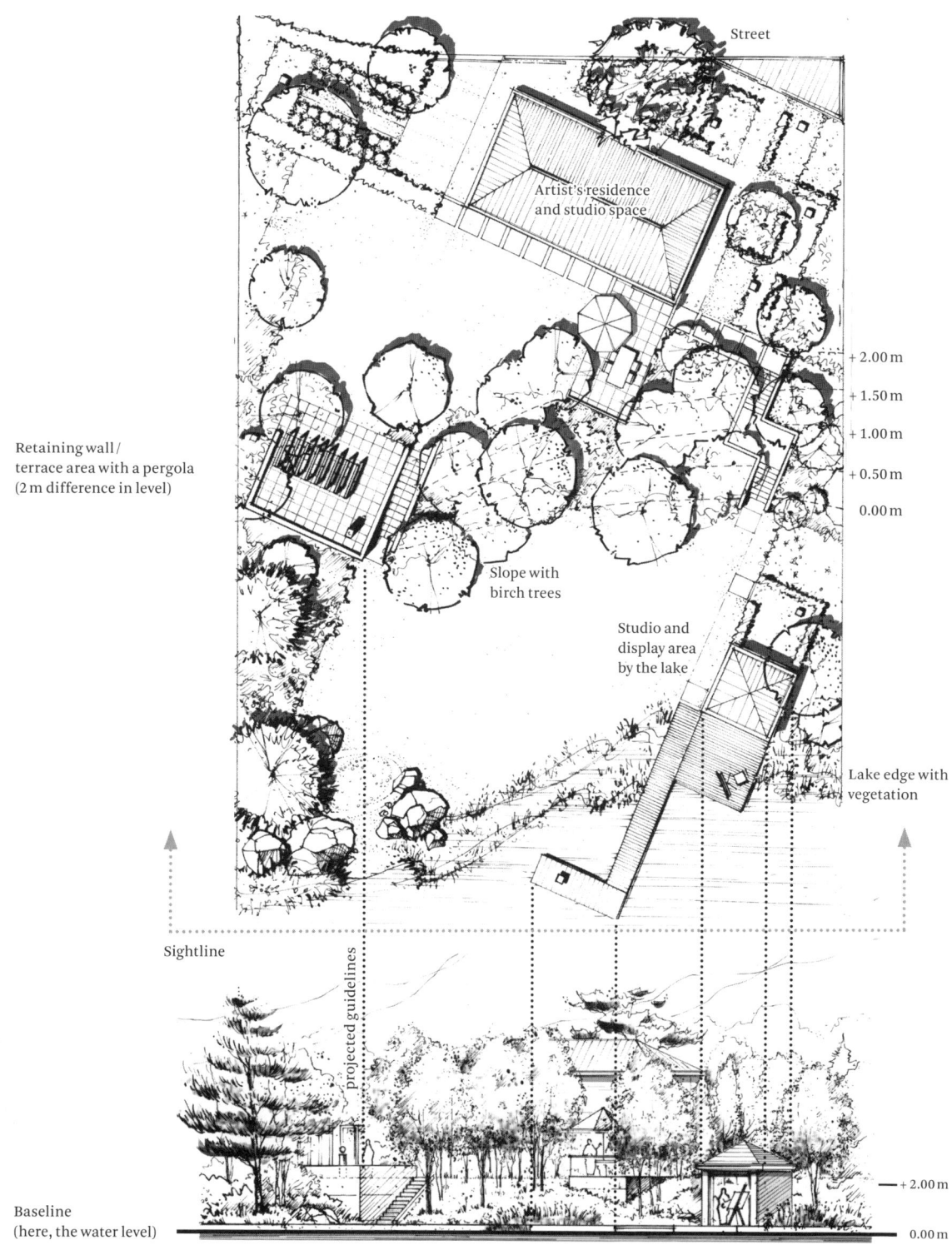

Elevation and section

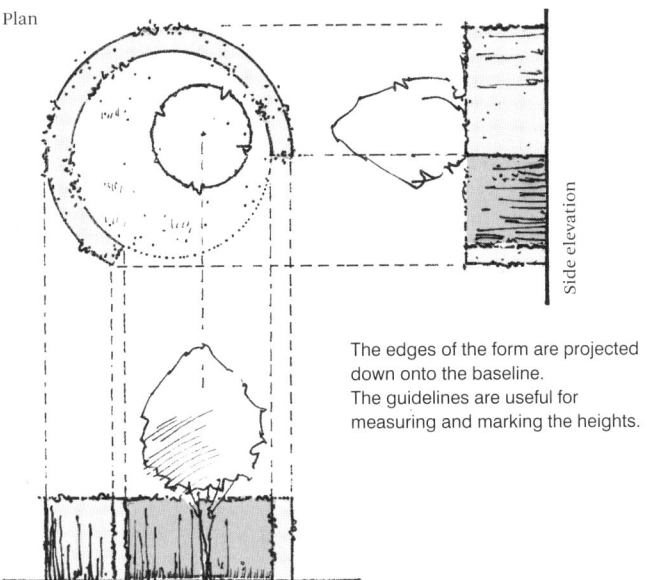

Plan

Side elevation

Frontal elevation

The edges of the form are projected down onto the baseline.
The guidelines are useful for measuring and marking the heights.

The hedges in elevation appear very flat because both the foreground and the background elements are projected to scale.

The perspective will always communicate a spatial experience far more effectively than a projected elevation. However, a perspective is not to scale.

In stark contrast to a perspective, the elevation projects all elements in the foreground and background to scale onto the picture plane. There is no foreshortening and all shapes retain their true proportions and dimensions. The elevation may not be as visually attractive as a perspective; however, it has the advantage of being to scale. With the help of graphic techniques, such as shadows, textures and contrasts, an elevation can be rendered spatially even if it retains a slightly abstract feel.

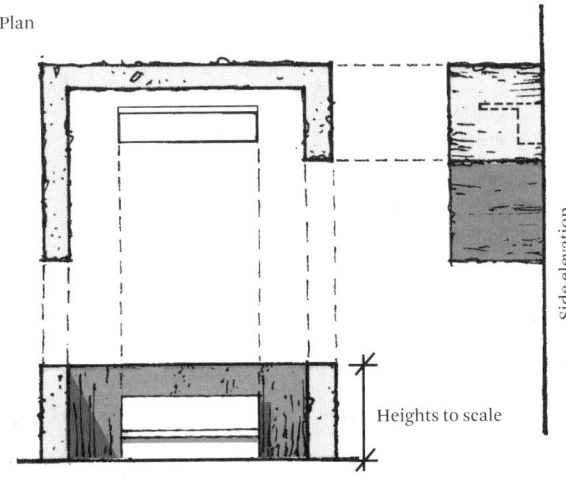

Plan

Side elevation

Heights to scale

Frontal elevation

114

Elevation
Introduction
Construction

4

Plan and elevation work together to communicate different aspects of a design. In the examples on the last page, the geometric shapes of the hedges were only visible in the plan view and lost in the elevation. In contrast, the elevation on this page reveals the wavy hedges much better than the plan.

Elevation and section

Trees in elevation
A tree symbol in a plan shows exactly where the tree will be located in the space. Although the irregular and loose outline already relays some information about the tree to be planted, an elevation reveals much more than the plan ever can. Trees are an important part of an elevation, both in architecture and landscape architecture. Every tree type has its own characteristic growth pattern, typical form or outline (its so-called *Habitus*). In order to properly sketch specific tree types, it is a good idea to observe real trees and practise sketching their distinct shapes and their characteristics in a sketchbook.

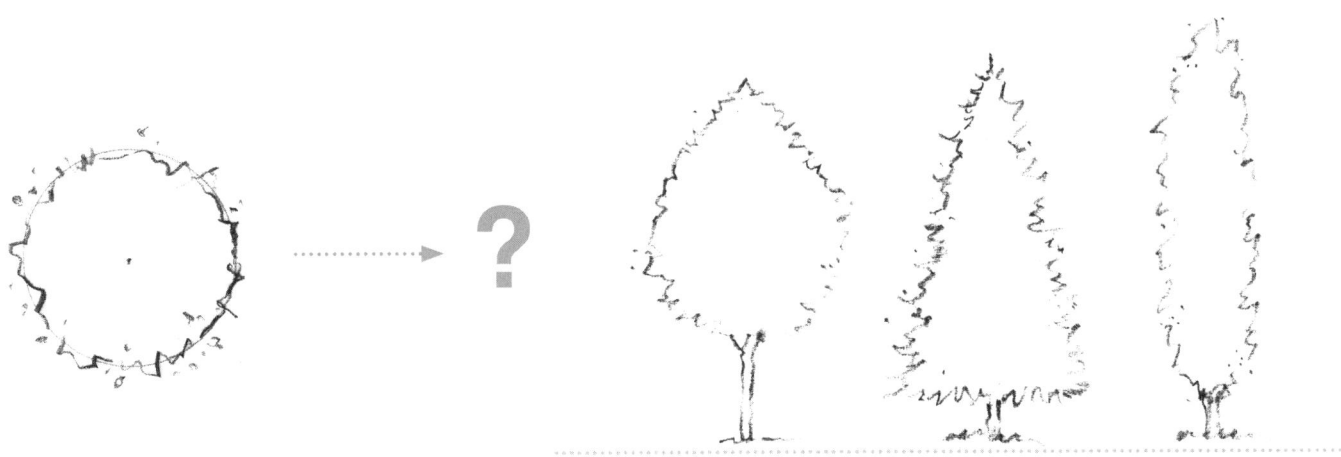

Vegetation in elevation
Trees
Shrubs and woody plants
Potted plants
Ground cover, grasses, and flowering plants
Adding depth

4

Drawing simple tree outlines is a good way to get a feeling for the trees form and proportions, even if they will need to be reduced in detail. Sketching trees can also assist when learning tree names and their Latin nomenclature. The better we know our trees, the easier it is to draw them and to test ideas and design variations with them on paper.

Elevation and section

Foliage texture
The pure outline of a tree is a great starting point, however it can be enhanced with foliage texture which further expresses the tree character. With the help of a light source and texture gradients, the tree can receive graphic volume.

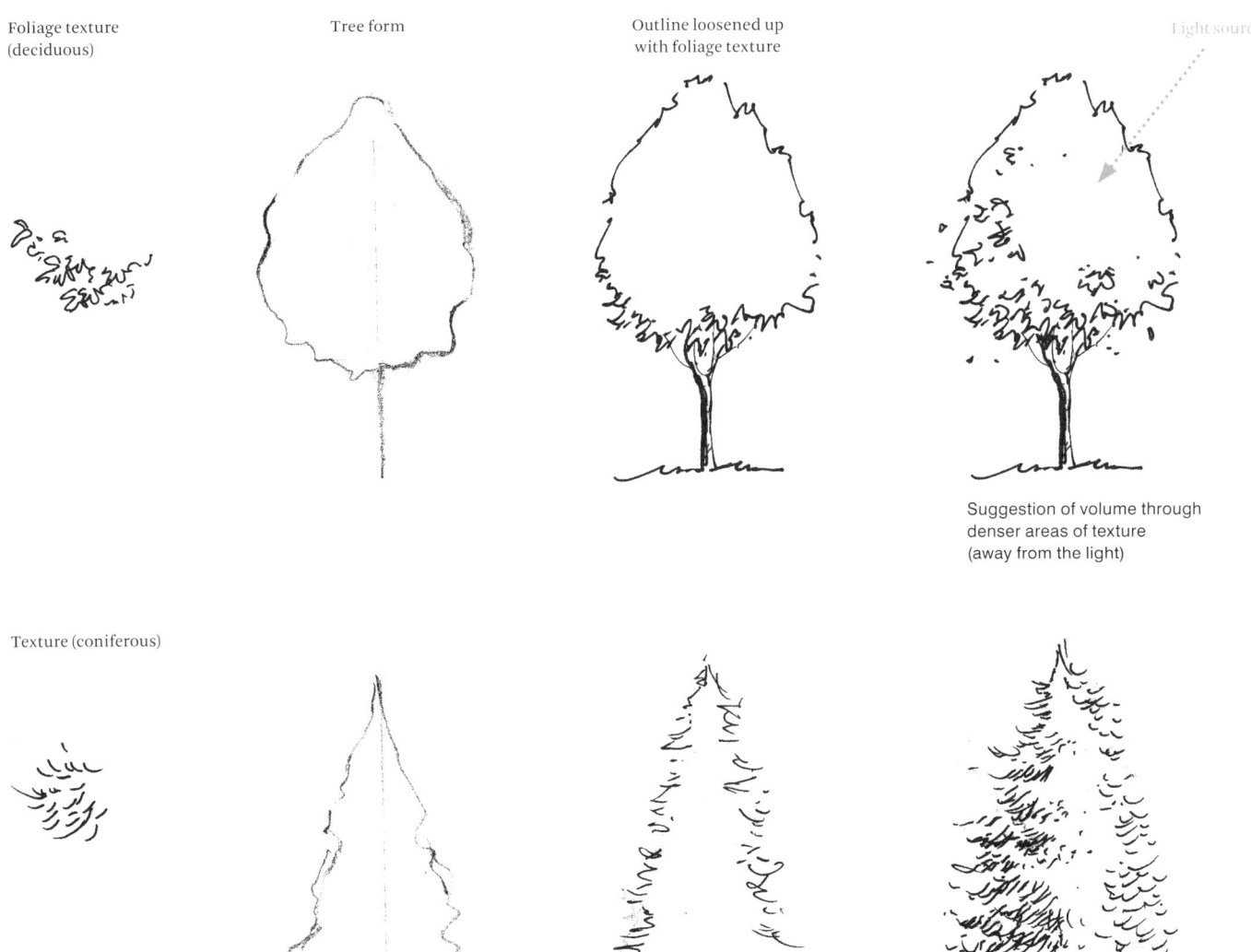

Vegetation in elevation
Trees
Shrubs and woody plants
Potted plants
Ground cover, grasses, and flowering plants
Adding depth

4

Degrees of abstraction

The degree of abstraction used in drawing trees can be varied, depending on the project phase, the time available and the viewer. Quicker trees will tend to have less small-scale textures and details. Quick and simple trees are great for the initial stages of a design, when ideas are being explored and it is not yet know which tree will be planted. Regardless of graphic style, the quality of an elevation will depend on the drawing style used and the consistency of this style throughout the overall composition.

The outline is always the starting point from which to communicate a specific tree type

Expressing light and shade in the tree crown is a great way to give the tree symbol volume

Branch pattern trees take longer to produce and have an interesting graphic quality, even if they do not reveal much about a specific tree type

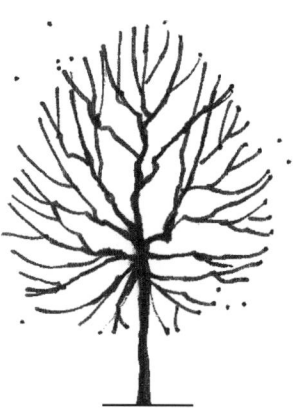 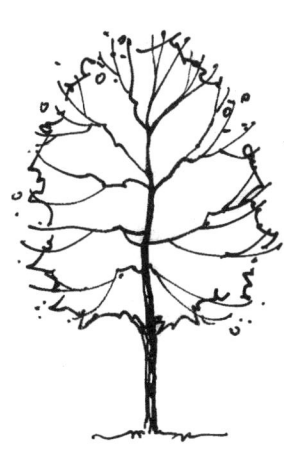

119

Elevation and section

Trees in elevation
Drawing trees will always include the potential for personal expression. These symbols can be very subjective interpretations of tree types and their characteristics.

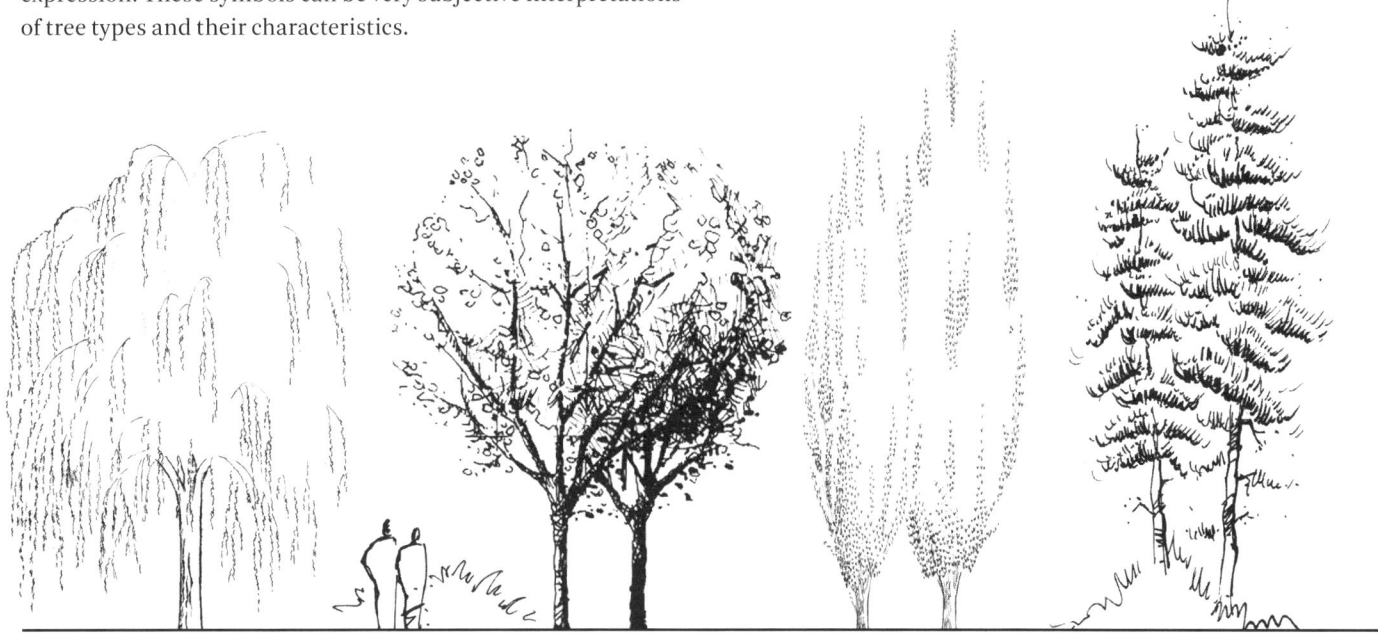

Vegetation in elevation
Trees
Shrubs and woody plants
Potted plants
Ground cover, grasses, and flowering plants
Adding depth

4

Overlapping elements can add a sense of depth to the scene. The heavier the delineation, the closer it appears to the viewer.

Elements in the foreground can thus be drawn using stronger line weights. Objects in the background can have much finer outlines and be rendered to a much lesser degree.

Elevation and section

More trees in elevation
As with the plan view, remember to contrast foreground and background when grouping trees in an elevation. Not everything needs to be rendered to an equally high degree and with the same line weight.

The quicker we sketch, the less time we have for details.

Vegetation in elevation
Trees
Shrubs and woody plants
Potted plants
Ground cover, grasses, and flowering plants
Adding depth

4

Aligning the trees in plan and elevation increases legibility. Graphic similarities also help viewers read and understand the information presented.

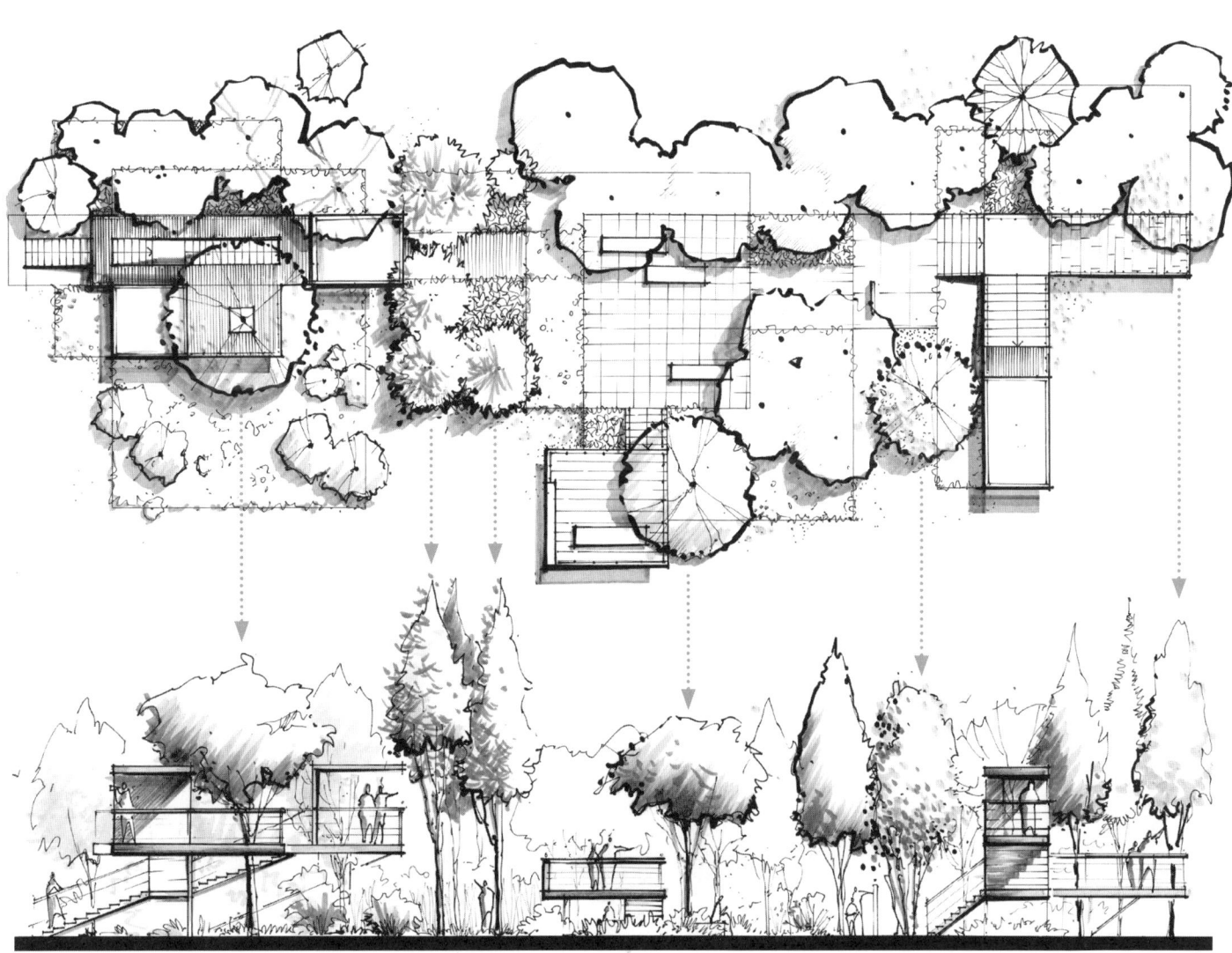

Elevation and section

More trees and shrubs in elevation
Stylistic abstractions are useful in the preliminary design phase, when discussion focuses on an idea and not necessarily on specific types of trees or shrubs.

Vegetation in elevation
Trees
Shrubs and woody plants
Potted plants
Ground cover, grasses, and flowering plants
Adding depth

4

The site plan and elevation are almost always presented together and should have a legible reference to each other in the layout.

The elevation shown here (and also the one on page 127) refers to the site plan, projects the 'flat' tree symbols vertically downwards and shows the exact tree species at a measurable height. The planned contents are immediately clear.

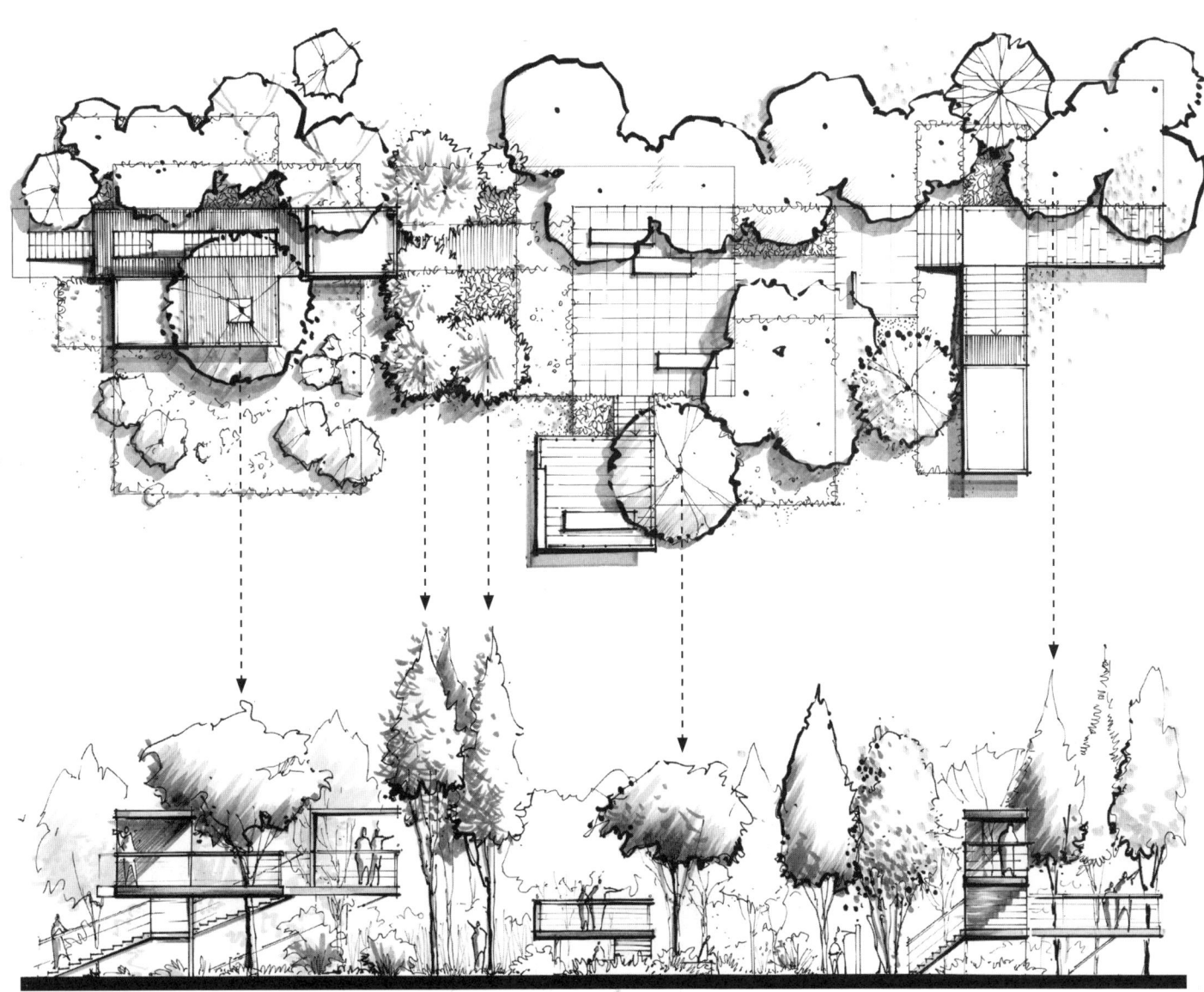

Elevation and section

Quick trees
These trees are quick and easy to sketch, some possessing a very expressive quality.

Vegetation in elevation
Trees
Shrubs and woody plants
Potted plants
Ground cover, grasses, and flowering plants
Adding depth

4

When plan and elevation are presented together, it is important to retain graphic continuity. Styles and lines should appear similar in both projections.

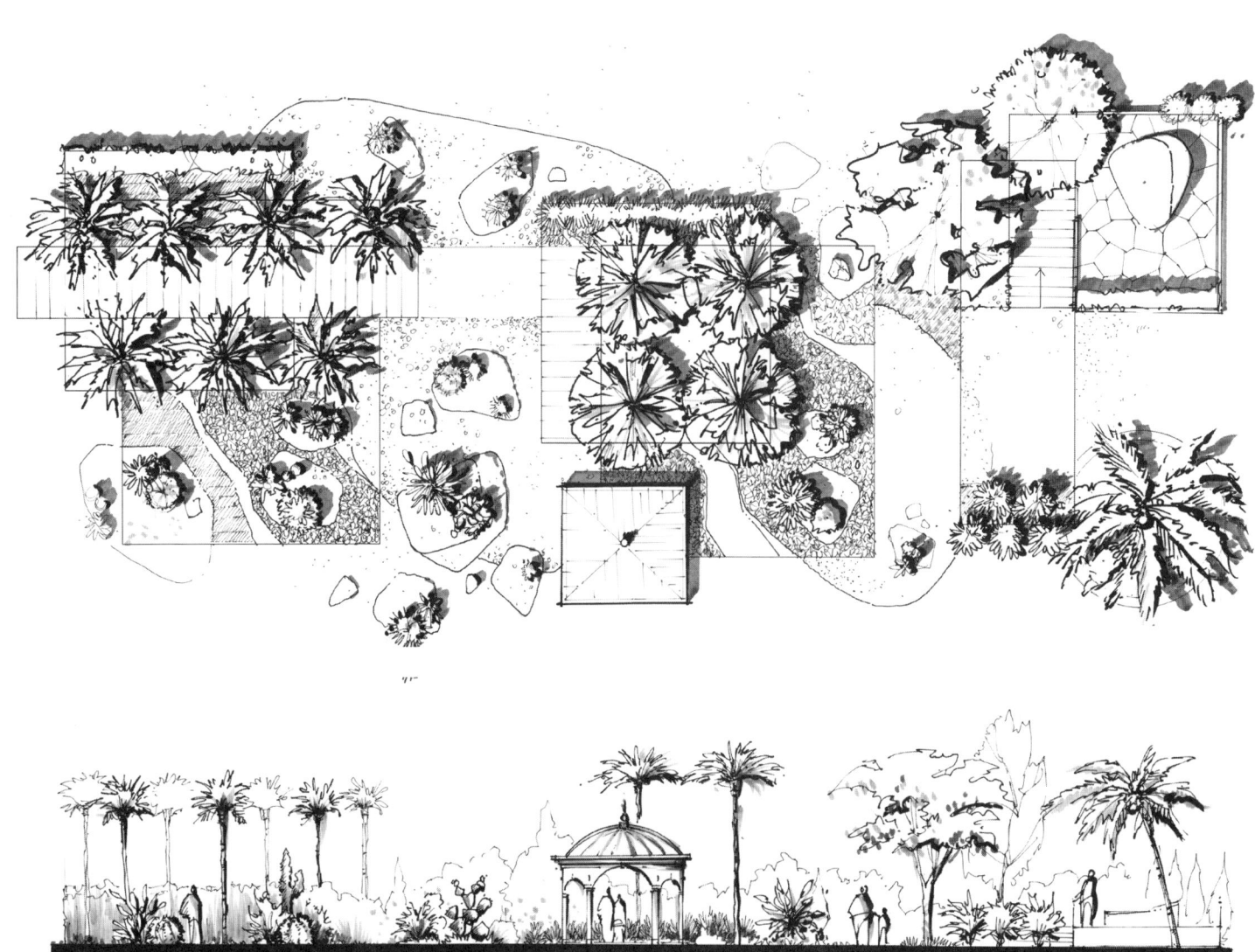

Elevation and section

Shrubs and woody plants
Shrubs and woody plants can be drawn in various degrees of detail and volume, depending upon their importance in an elevation. Overlapping them conveys a sense of depth (*see page 132*).

Vegetation in elevation
Trees
Shrubs and woody plants
Potted plants
Ground cover, grasses, and flowering plants
Adding depth

4

Clipped shrubs, woody plants, and topiary

Potted plants

129

Elevation and section

Ground cover, grasses, and flowering plants

Vegetation in elevation
Trees
Shrubs and woody plants
Potted plants
Ground cover, grasses, and flowering plants
Adding depth

4

Drawing individual and equidistantly spaced flowers is not a convincing way to convey them in elevation. Instead, draw the plant volumes using varied and irregular outlines, and with foliage textures that correspond to leaves of the actual plant. These volumes can be drawn overlapping each other, with blossoms and flowering elements emerging sparingly and irregularly.

Elevation and section

Adding depth
In an elevation, a sense of depth can occur by simply overlapping elements in the foreground and background. It is also strengthened through differentiation of line weights, as well as through different levels of rendering.

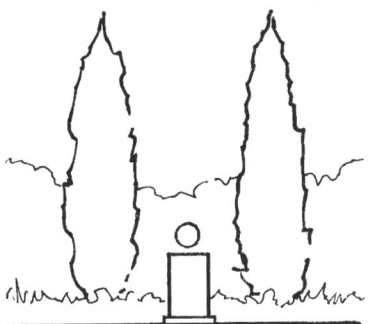
Graphic emphasis given to foreground through stronger line weights

Graphic emphasis on foreground through rendering

Equal emphasis foreground and background

Vegetation in elevation
Trees
Shrubs and woody plants
Potted plants
Ground cover, grasses, and flowering plants
Adding depth

4

The overlapping planes of an elevation and the different degrees of detail can create interesting scenes. They appear much like a stage set, where the foreground is highlighted and areas further back are merely hinted at.

Elevation and section

Built structures: Pergolas, pavilions, and arbours
The pergolas, garden pavilions, gazebos, and arbours which grace garden designs can take on many forms. Here are just a few examples for reference.

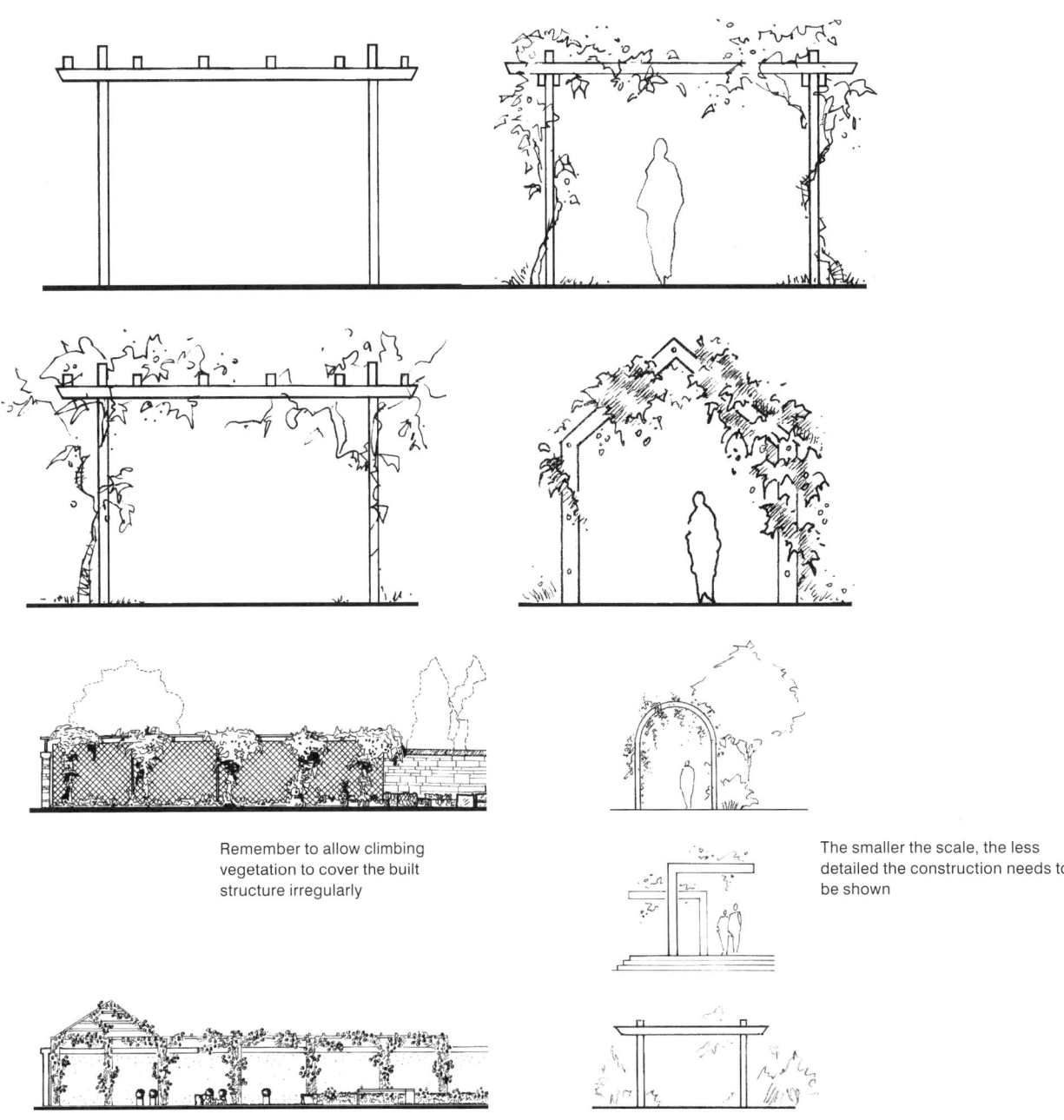

Remember to allow climbing vegetation to cover the built structure irregularly

The smaller the scale, the less detailed the construction needs to be shown

Built structures
Pergolas, pavilions, and arbours
Walls and materials
Water
People

4

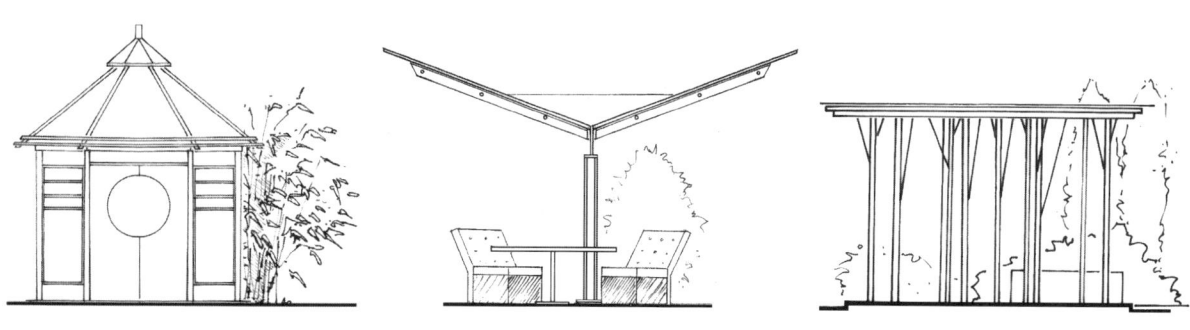

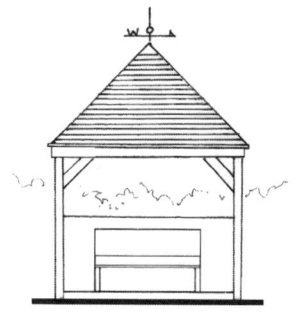
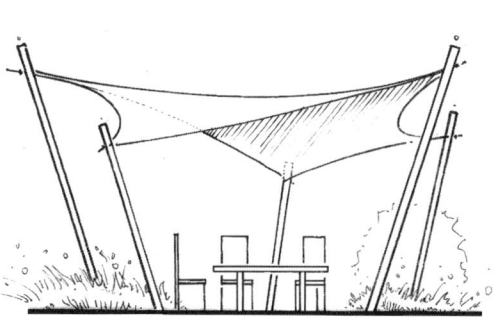
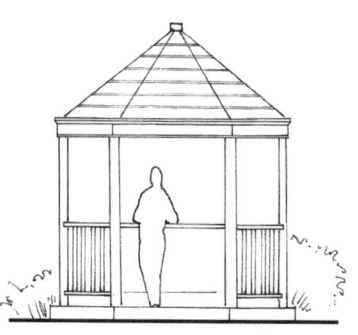

Elevation and section

Walls and surface materials
Even though CAD programs draw materials to a very precise degree, surfaces do still sometimes have to be sketched and indicated. How much detail is included depends upon the scale and the intention of the drawing.

If the sketch is not a construction drawing, it is usually sufficient to suggest the materiality, along with its patterns and distinct surface quality. At smaller scales, materials and their patterns are often highly simplified and reduced.

Brick

Dry stone wall

Natural stone blocks (regularly offset)

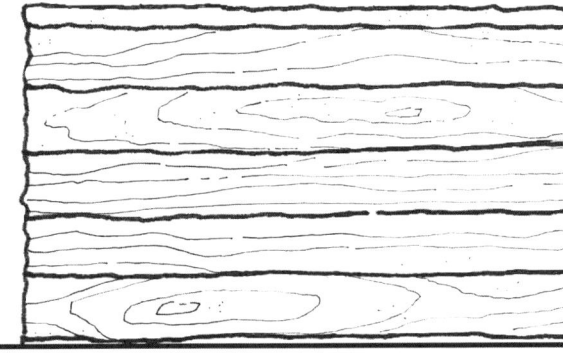
Wood

Natural stone blocks (irregular)

Built structures
Pergolas, pavilions, and arbours
Walls and materials
Water
People

4

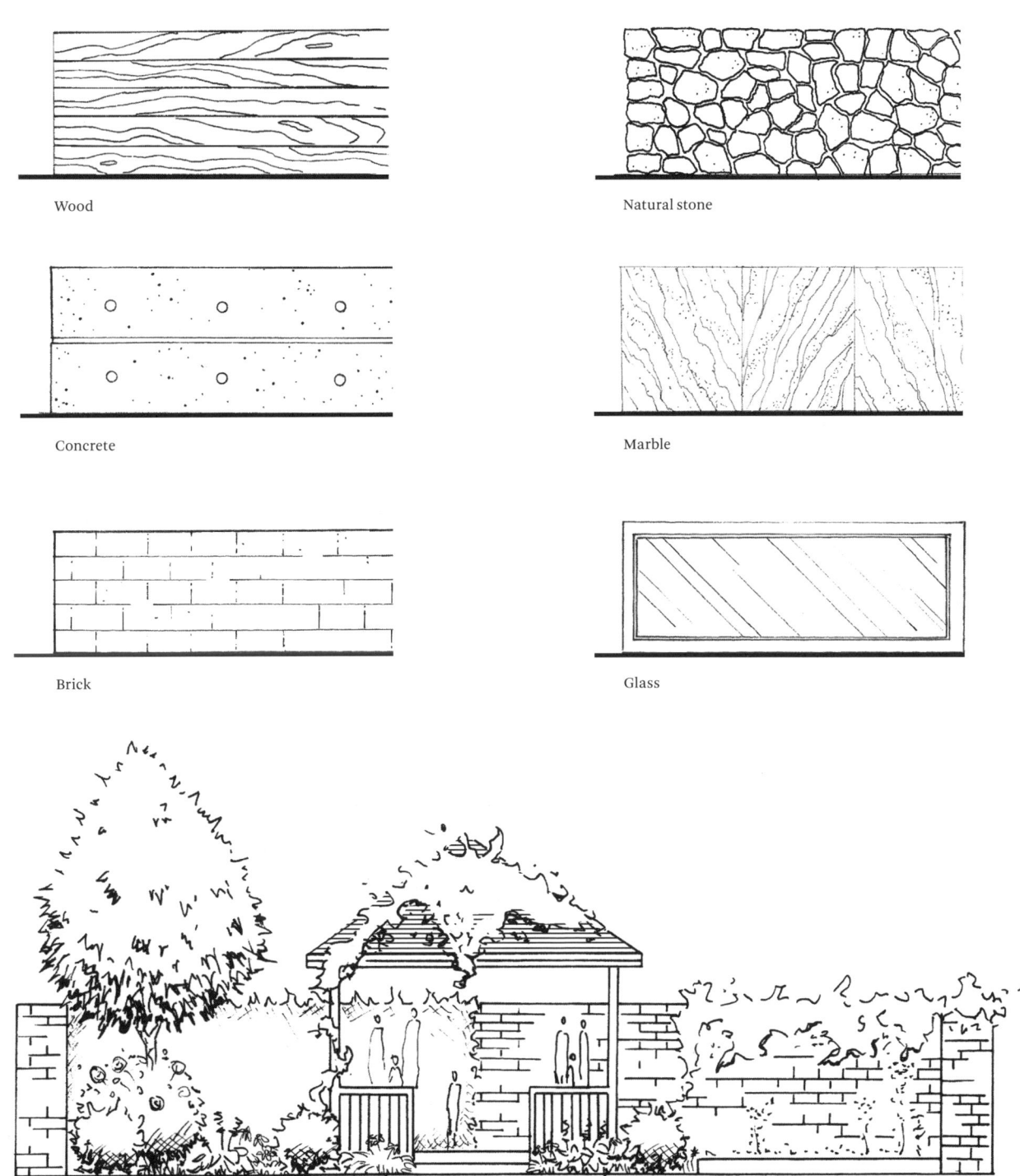

Wood

Natural stone

Concrete

Marble

Brick

Glass

Elevation and section

Water features in elevation
Similar to the techniques used in a plan view, falling or spraying water is drawn using fine dots and straight lines with irregular densities to indicate vertical movement.

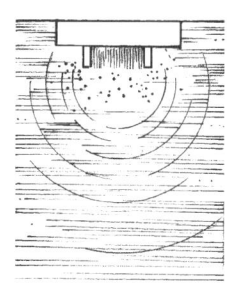

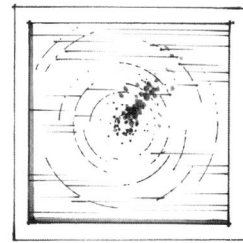

Plan view

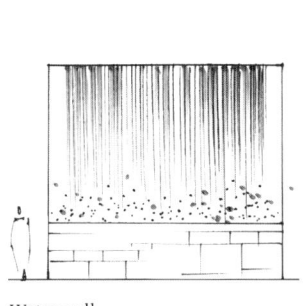

Water wall

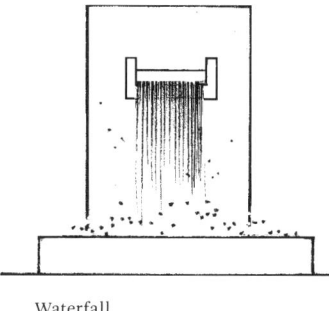

Waterfall

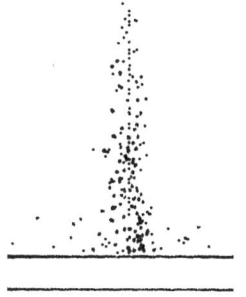

Fountain

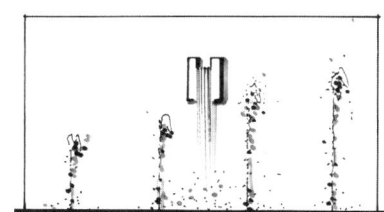

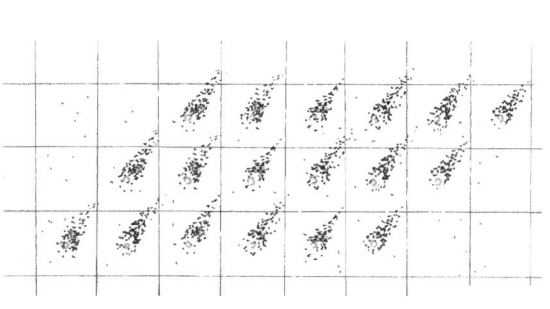

Water jets

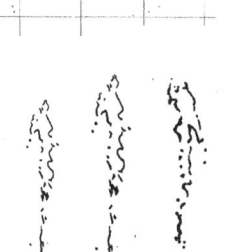

Water ball

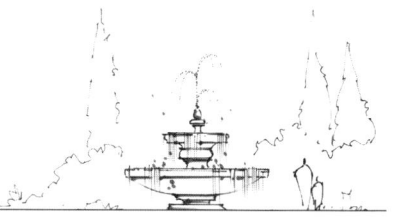

Built structures
Pergolas, pavilions, and arbours
Walls and materials
Water
People

4

People

Drawing people in sections and elevations is a great way to indicate scale, enliven a scene and show potential use of the space. Drawing people requires practise and a good knowledge of anatomy in order to get proportions right. In landscape architecture it is usually best to draw simple outlines of human figures without much detail. A small oval head, with shoulders and a slightly conical body shape towards the ground plane is sufficient for communicating the human form. Although it is great to show people in different poses and doing different things, too much detail might distract from the space itself, which should always be the focus of any architectural section or elevation.

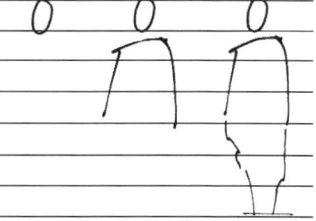

The classic head to body proportion of 1:7 can be used as a starting point for a sketch person

1:100

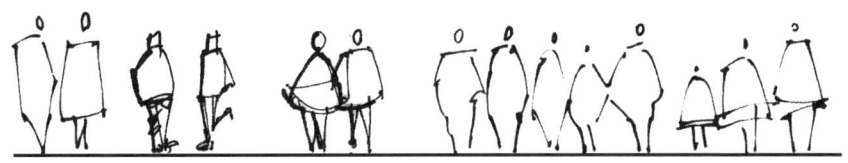

1:200

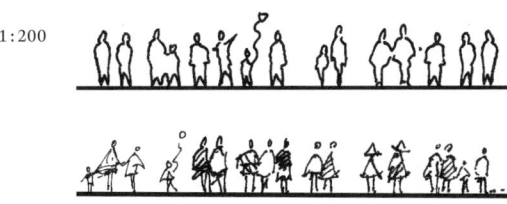

Elevation and section

Section
A section is a vertical cut through a building or a landform. When the pieces are separated, we are left with a cross section. The viewer looks at this cut area from a particular position and with a 90° angle. This is the true section. The section is graphically highlighted with a stronger outline or infilled area.

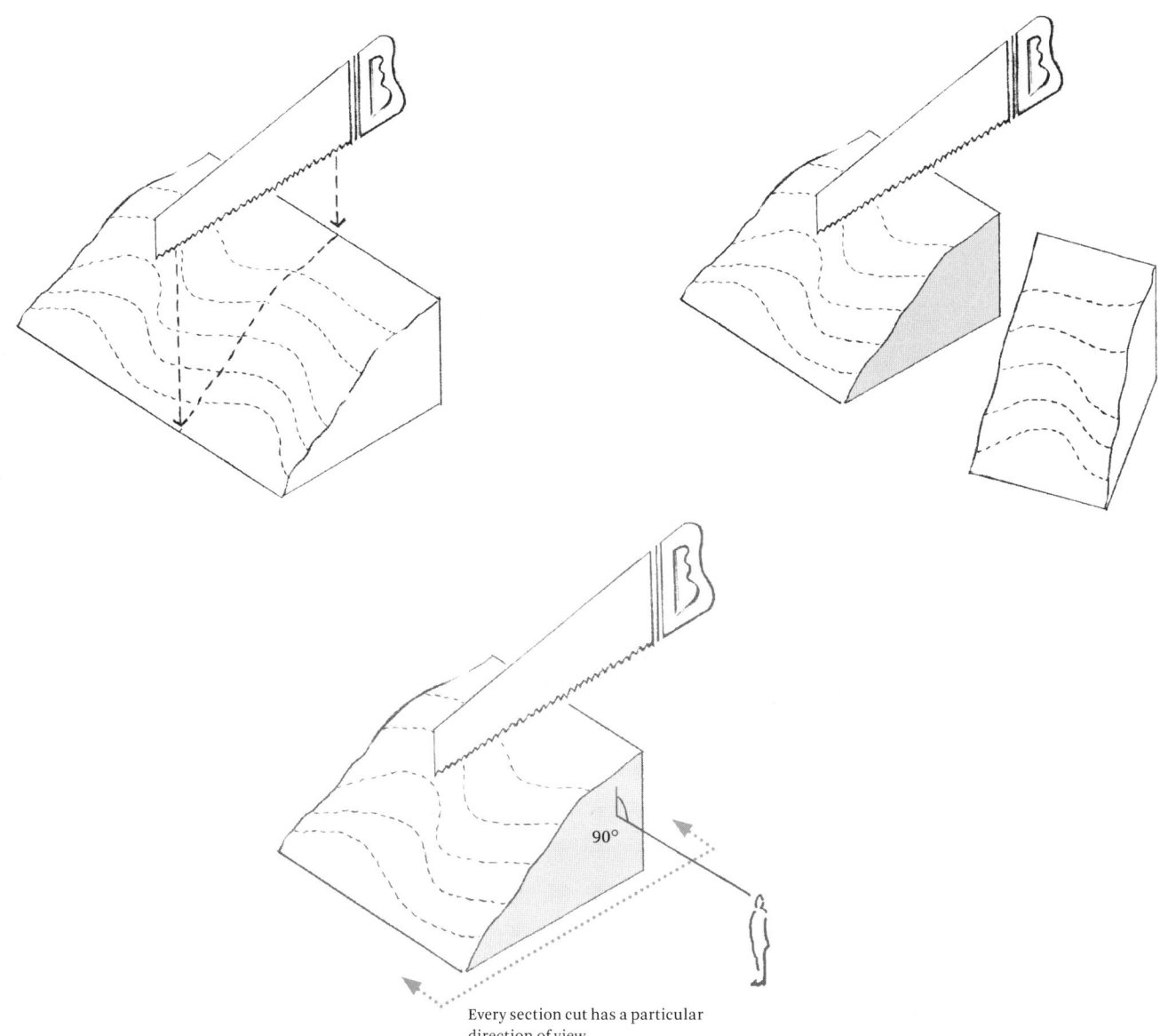

Every section cut has a particular direction of view

Section and section-elevation
Introduction
Sections through buildings
Section cut lines
Constructing a section
Uses and scales
Section cut area
Examples

4

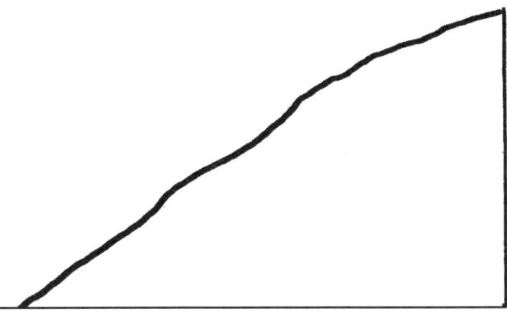

The true section reveals the level changes along the vertical surface of the landform

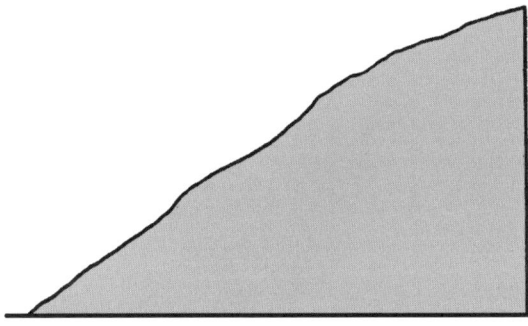

To increase legibility, the cut area is infilled with a tonal value

If the elevation area just behind the cut line is included in the view, it is referred to as a section-elevation. This is is much more graphically presentable and is very useful to landscape architects.

Elevation and section

Sections through buildings
Section cuts through buildings show the relationships of interior and exterior spaces. These sections communicate the main elements of the building, such as window openings, walls, roofs, columns, which are all made visible at once. The building's relationship to topography and its vertical changes in terrain are also shown.

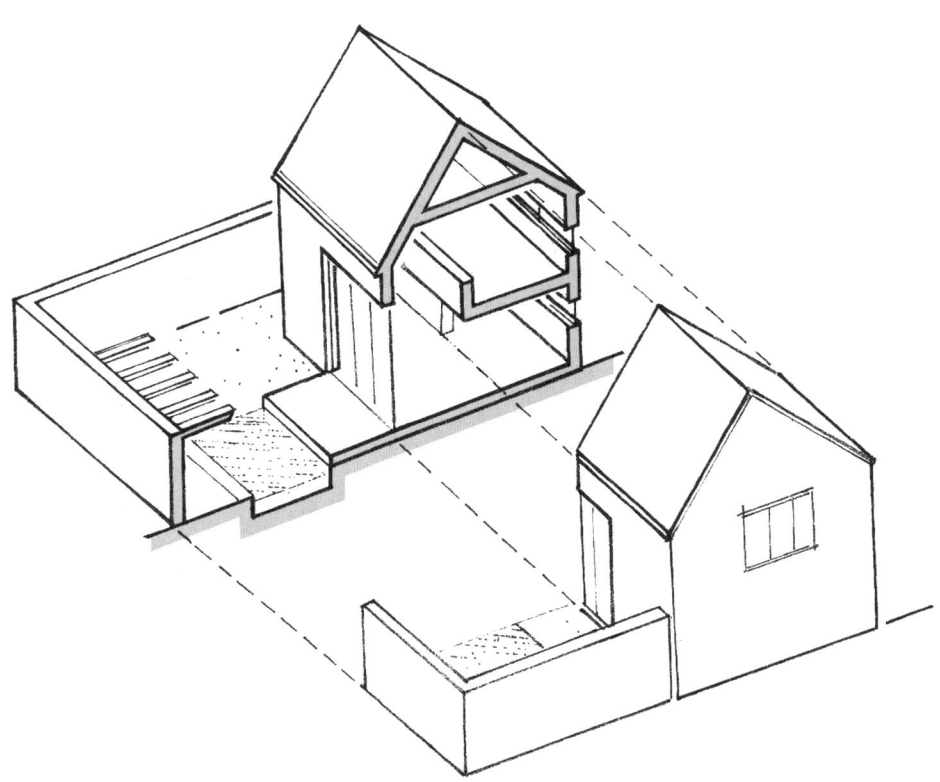

Section and section-elevation
Introduction
Sections through buildings
Section cut lines
Constructing a section
Uses and scales
Section cut area
Examples

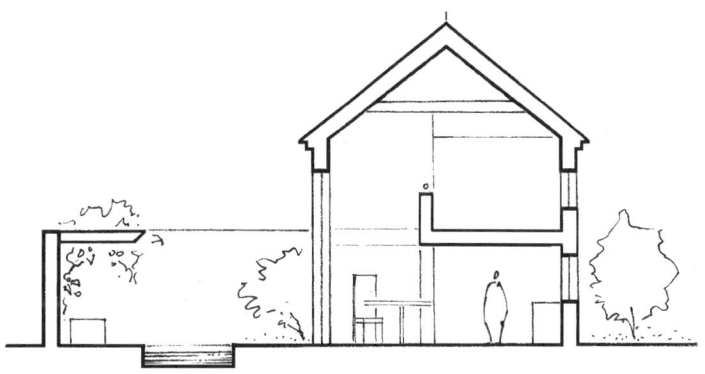

The cut surface should be indicated first and foremost with a stronger line weight. This can then be infilled with tonal values as necessary.

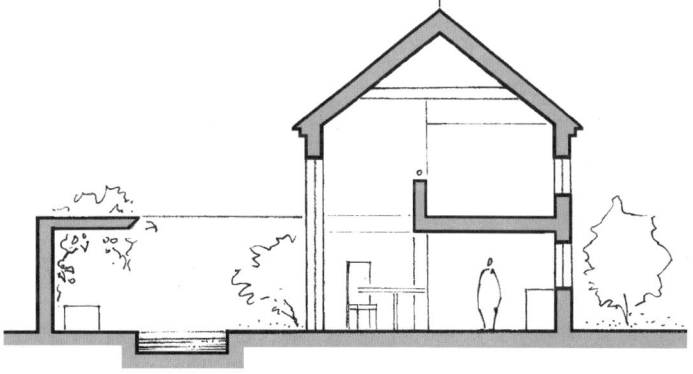

The strongest contrast and graphic effect come when the cut surfaces, here the walls, are filled with solid black

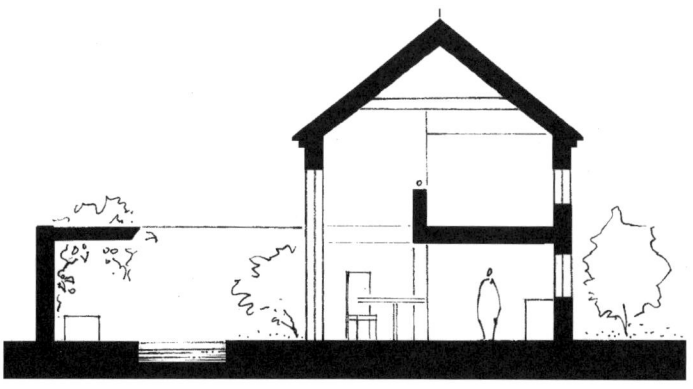

Elevation and section

Section cut lines

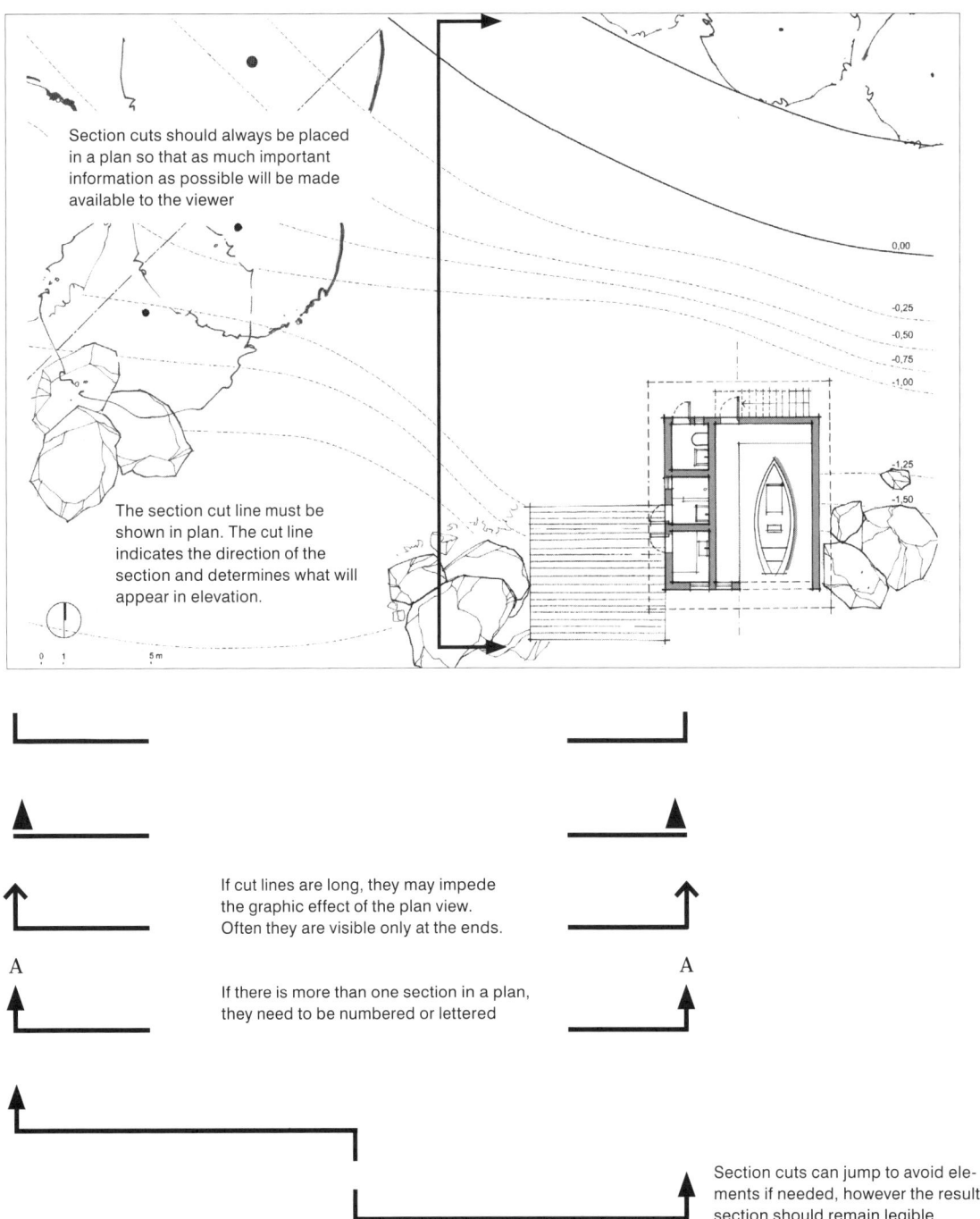

Section cuts should always be placed in a plan so that as much important information as possible will be made available to the viewer

The section cut line must be shown in plan. The cut line indicates the direction of the section and determines what will appear in elevation.

If cut lines are long, they may impede the graphic effect of the plan view. Often they are visible only at the ends.

If there is more than one section in a plan, they need to be numbered or lettered

Section cuts can jump to avoid elements if needed, however the resulting section should remain legible

Section and section-elevation
Introduction
Sections through buildings
Section cut lines
Constructing a section
Uses and scales
Section cut area
Examples

4

Constructing a section

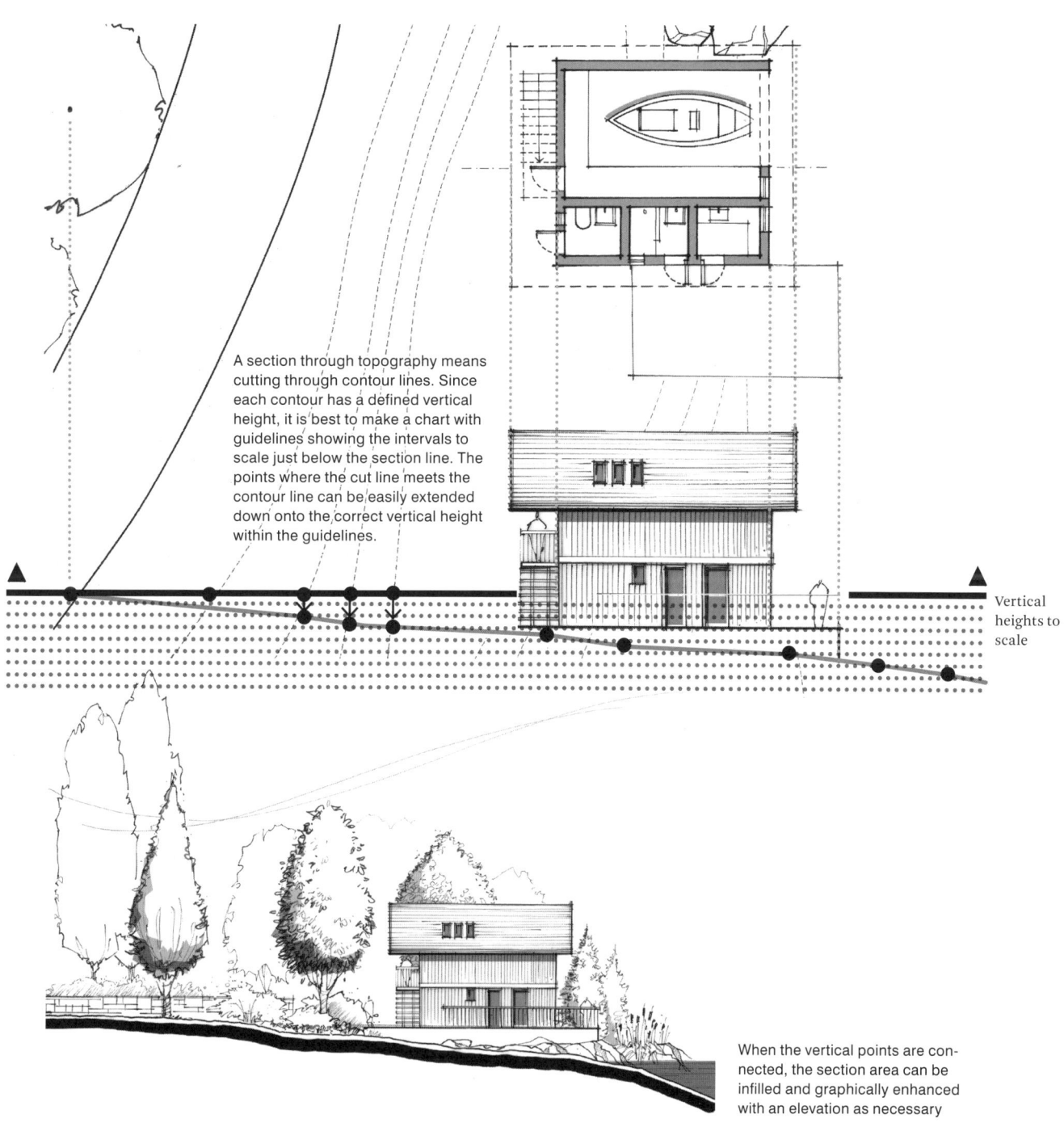

A section through topography means cutting through contour lines. Since each contour has a defined vertical height, it is best to make a chart with guidelines showing the intervals to scale just below the section line. The points where the cut line meets the contour line can be easily extended down onto the correct vertical height within the guidelines.

Vertical heights to scale

When the vertical points are connected, the section area can be infilled and graphically enhanced with an elevation as necessary

Elevation and section

Different uses and scales

1:100

Sections can be used in every scale and have many different applications

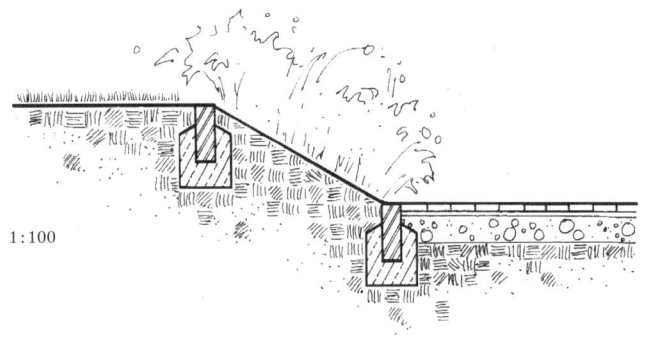

They can give detailed information about construction and materials

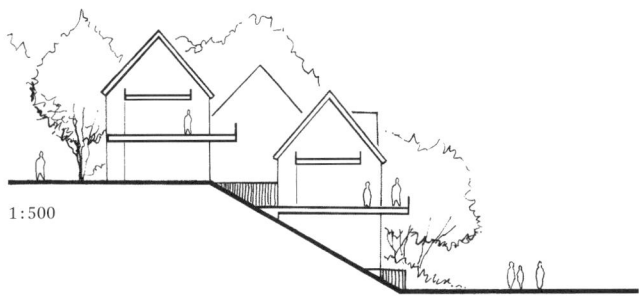

1:500

or show the positioning of buildings within topography

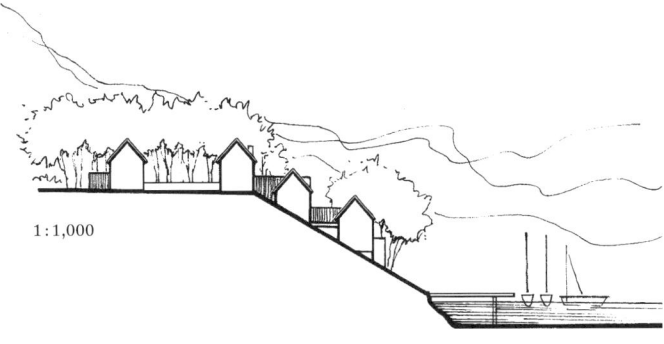

1:1,000

and describe the overall relationship between the built environment and its natural surroundings

Section and section-elevation
Introduction
Sections through buildings
Section cut lines
Constructing a section
Uses and scales
Section cut area
Examples

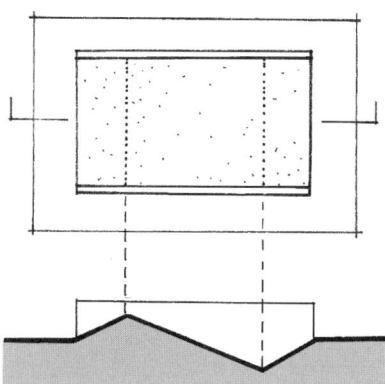

Earth moving

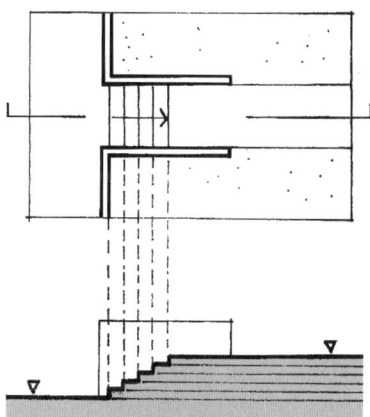

Stairs

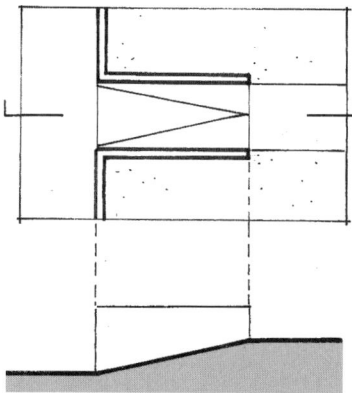

Ramps

Sections communicate topography and key level changes in a more legible way compared to a plan view. They give important, if highly abstracted, information that we would not otherwise see.

Elevation and section

Section cut area

There are many ways to graphically enhance a section cut area. A thicker line weight immediate distinguishes the cut area.

A solid black section cut area provides the strongest contrast and can easily be read. As this contrast also carries a heavy graphic weight, it might be preferable to fill the cut area with a grey tonal value instead.

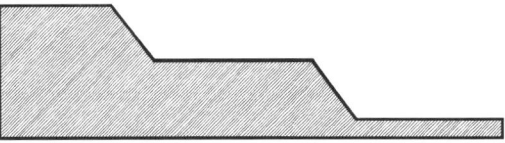

Quick freehand sections may be left without infill; in this case one could adopt a thick line to describe the surface and its level changes.

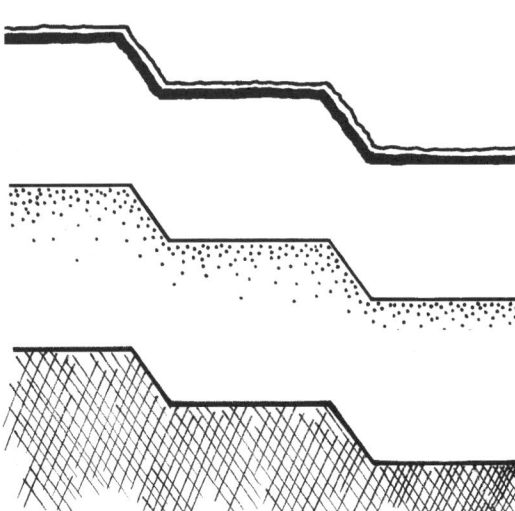

When filling the section cut area, it is best to finish the lower area with a straight horizontal line, otherwise it may compete with the upper (and more important) cut line.

Section and section-elevation
Introduction
Sections through buildings
Section cut lines
Constructing a section
Uses and scales
Section cut area
Examples

4

These examples show various ways to draw and render a section

The section cut has an important graphic presence in the overall composition which should be carefully considered

149

Elevation and section

More examples

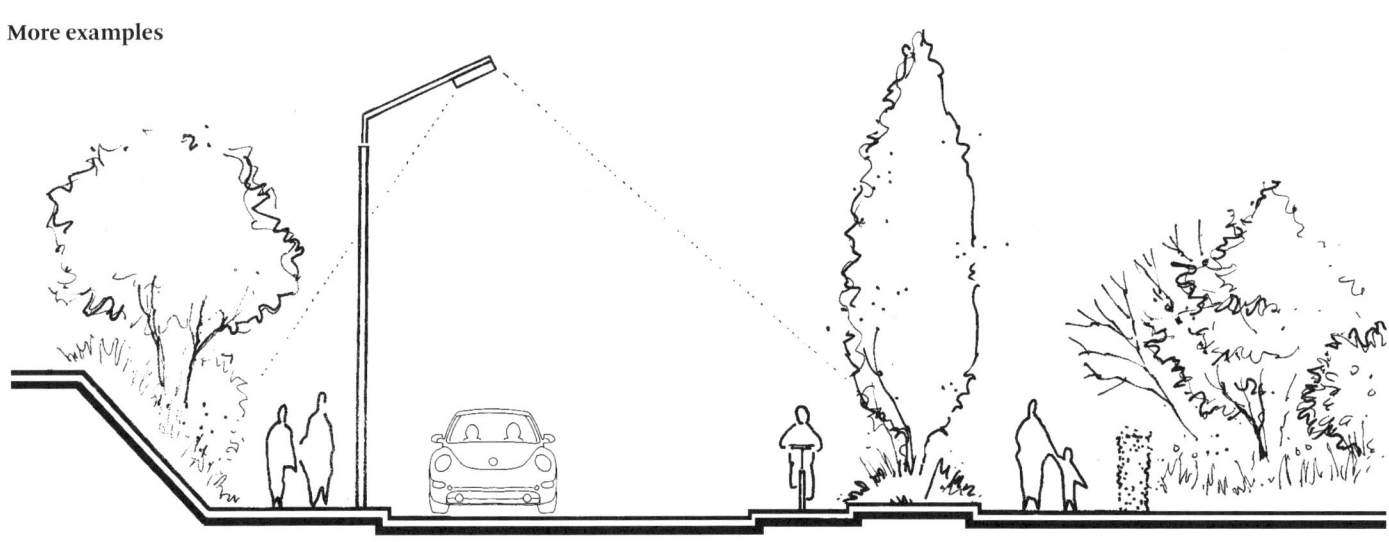

A section through a street area can show its different uses and functions

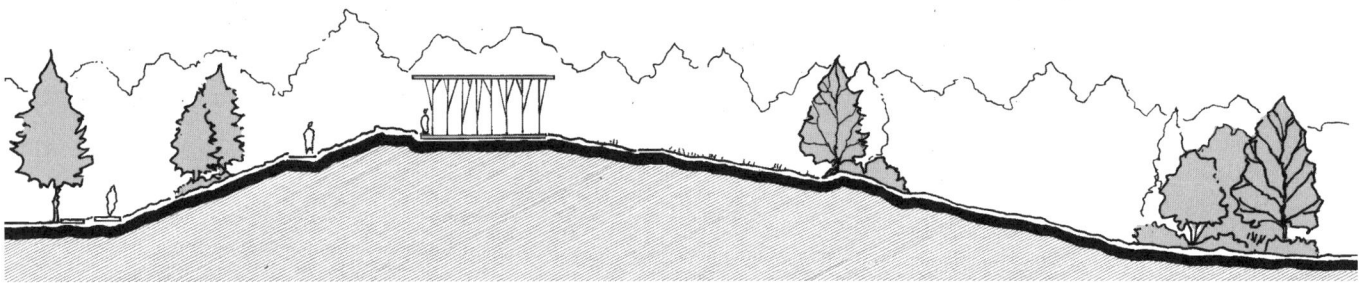

Sections through landforms highlight topographical level changes and height differences that we might not normally experience

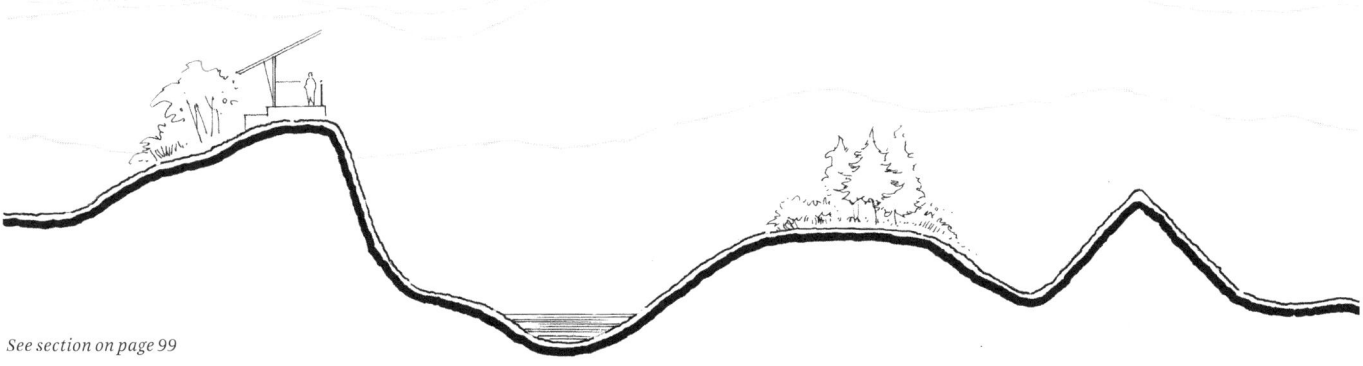

See section on page 99

Section and section-elevation
Introduction
Sections through buildings
Section cut lines
Constructing a section
Uses and scales
Section cut area
Examples

4

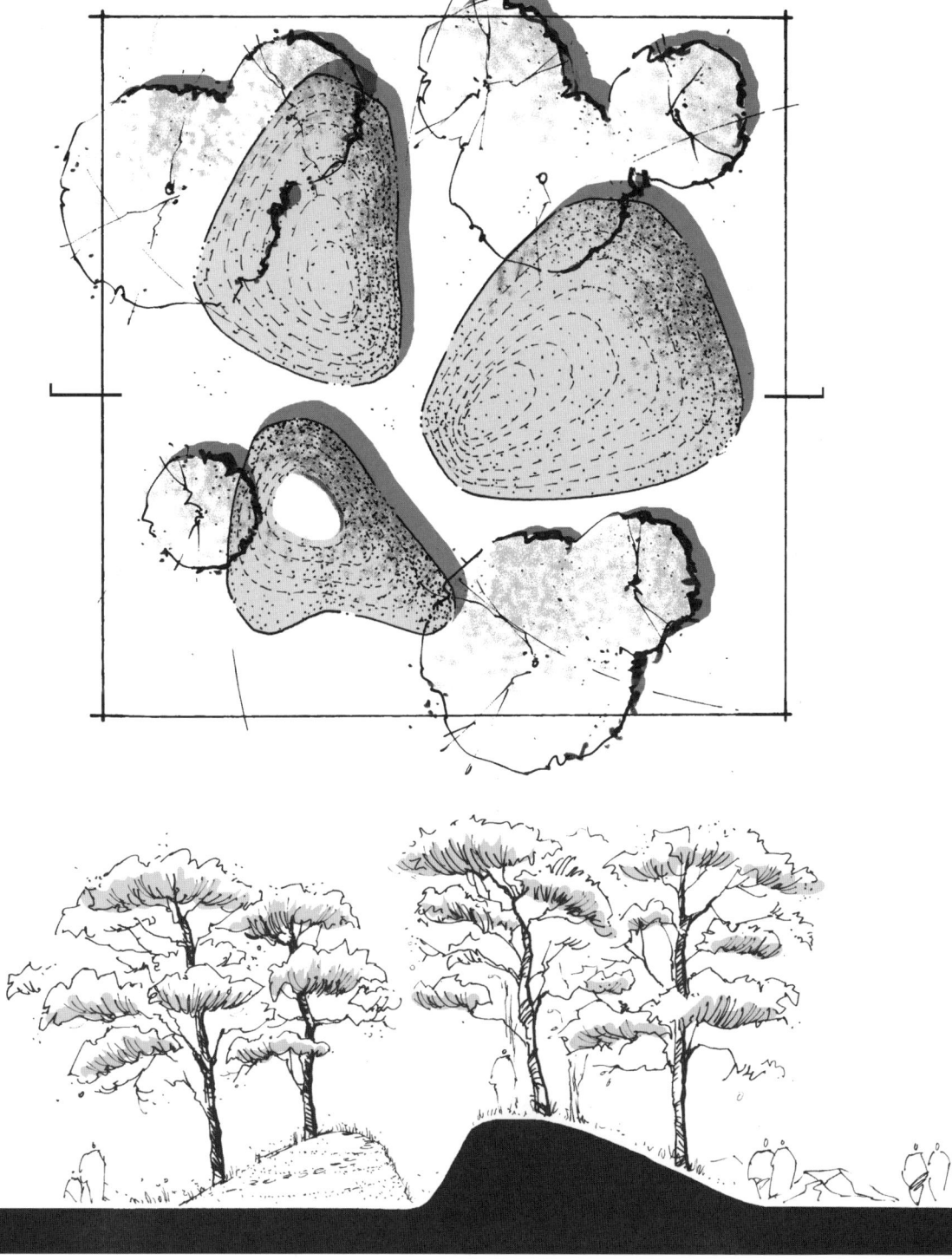

Elevation and section

Sections can make those elements of a design project visible, which would remain hidden in a plan view.

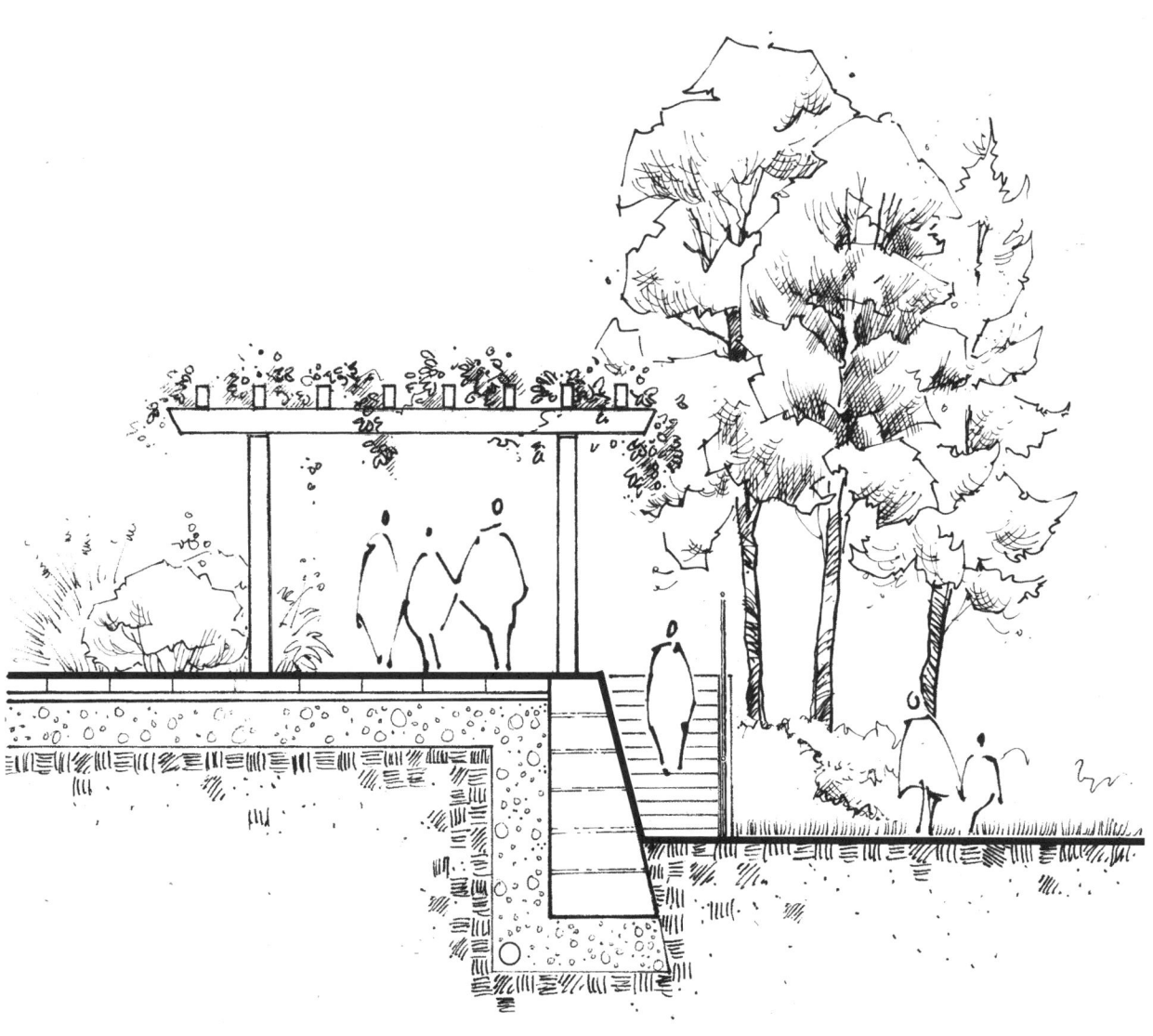

Section and section-elevation
Introduction
Sections through buildings
Section cut lines
Constructing a section
Uses and scales
Section cut area
Examples

4

There are numerous graphic possibilities to consider when drawing sections. Even simple sections can be graphically interesting.

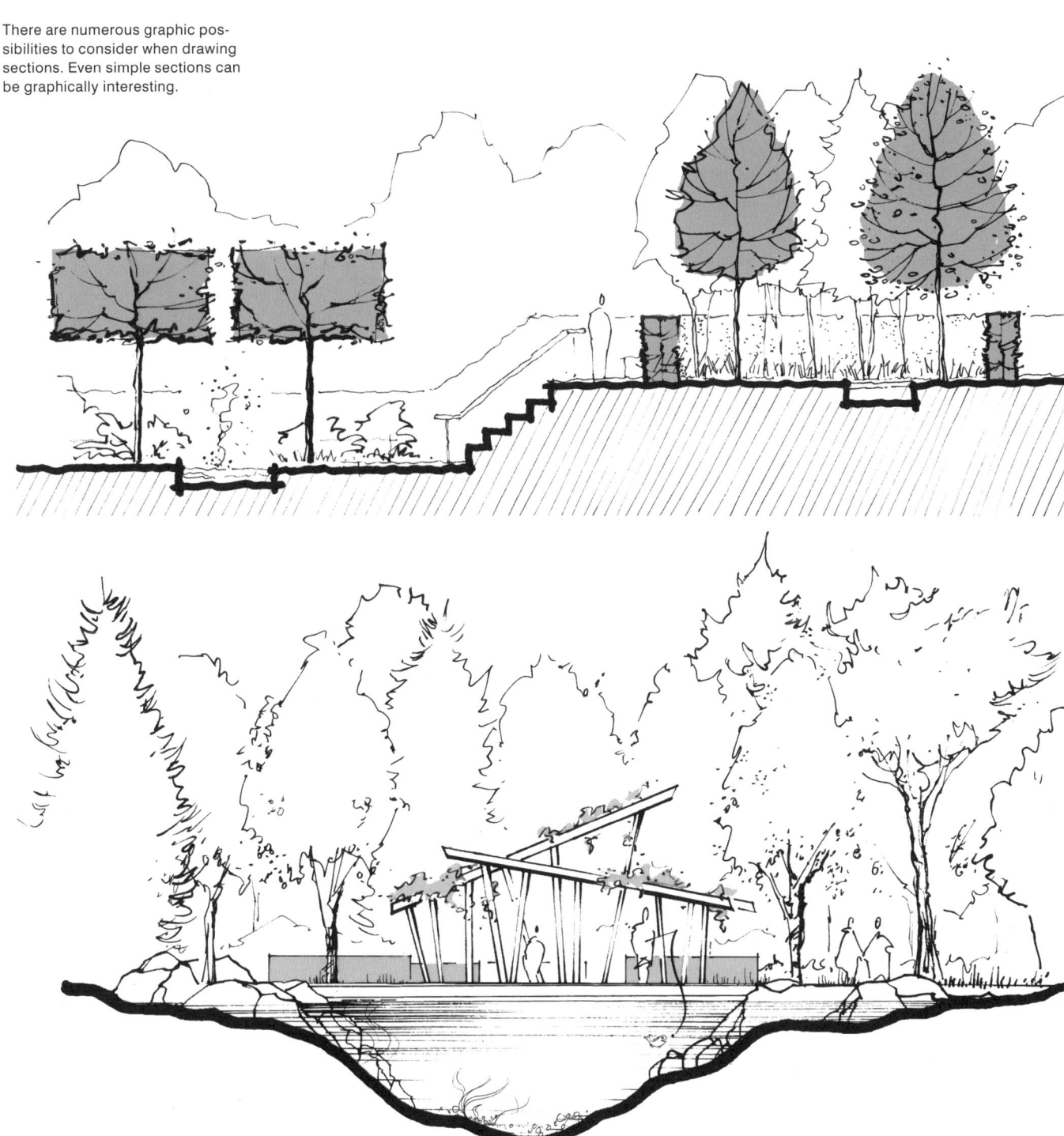

Elevation and section

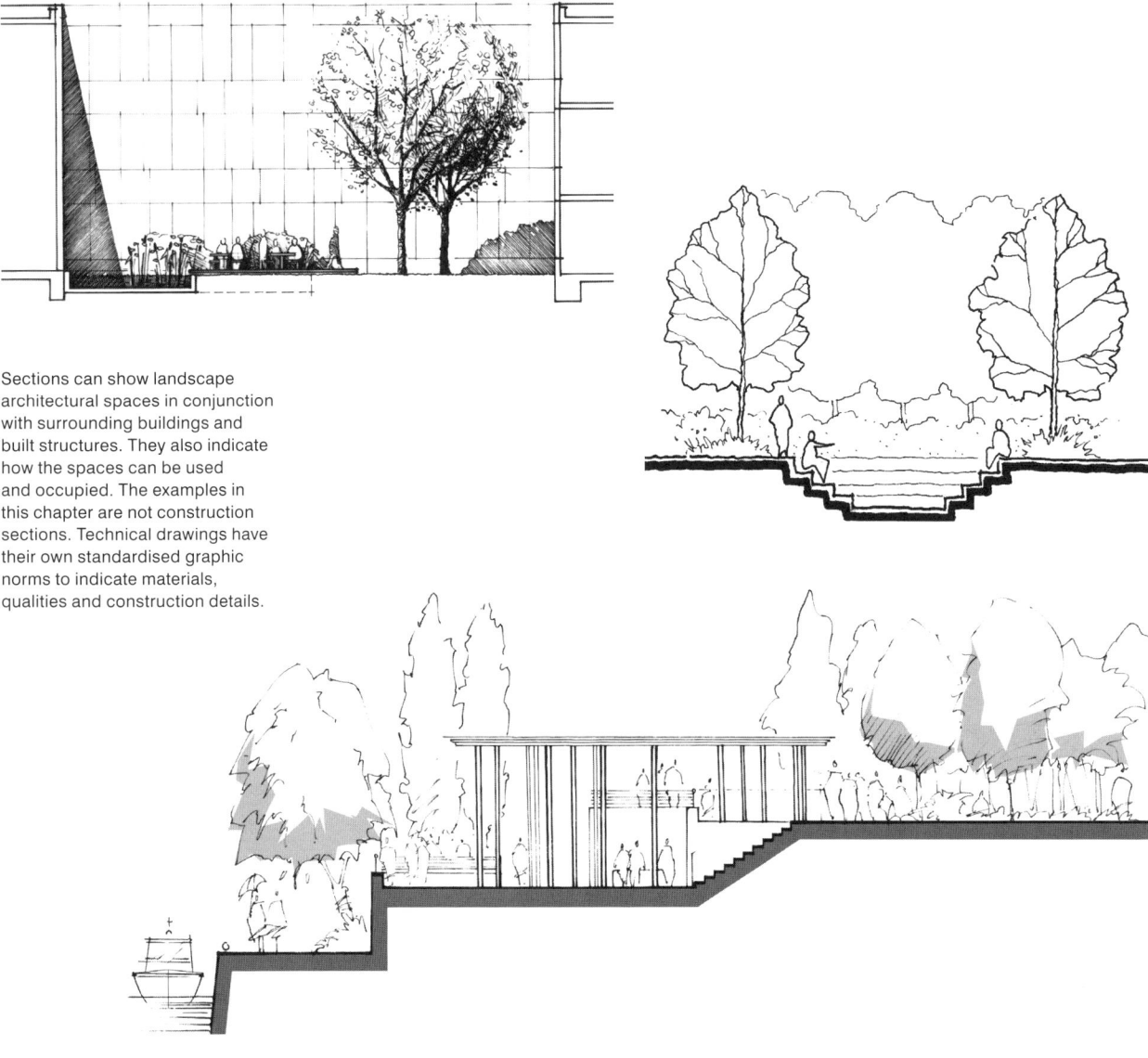

Sections can show landscape architectural spaces in conjunction with surrounding buildings and built structures. They also indicate how the spaces can be used and occupied. The examples in this chapter are not construction sections. Technical drawings have their own standardised graphic norms to indicate materials, qualities and construction details.

Section and section-elevation
Introduction
Sections through buildings
Section cut lines
Constructing a section
Uses and scales
Section cut area
Examples

4

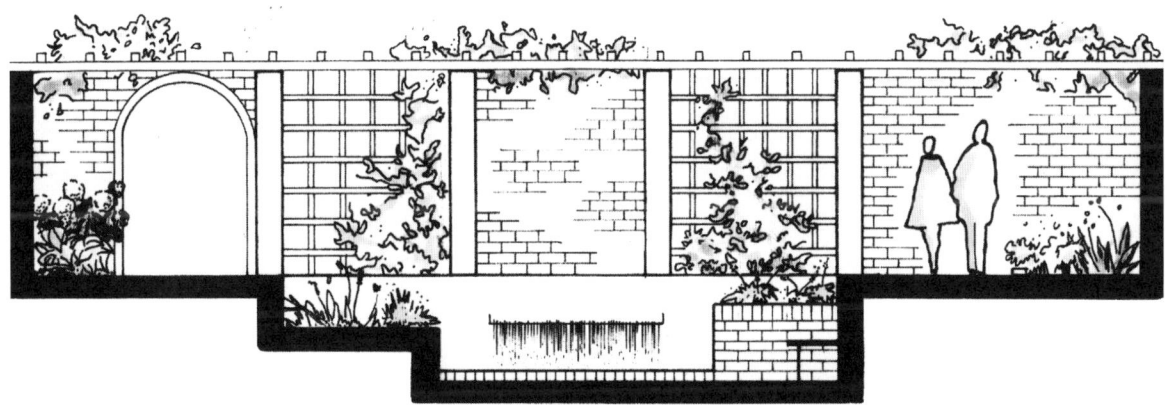

155

Parallel projections

Parallel projections
158 **Introduction**
159 **Isometric**
Elevation oblique
160 **Introduction and construction**
Axonometric projection
161 **Introduction and construction**
162 **Circles**
163 **Trees**
164 **Vegetation**
166 **Construction steps**

Parallel projections

Parallel projections: Introduction
Parallel projections, also called paraline views, are three-dimensional representations which are frequently found in architectural presentations. In these drawings, parallel lines are projected parallel along three different axes.
As the flat orthogonal projections only show two dimensions, the parallel projections have the advantage by showing three dimensions simultaneously. This means they are much more graphically effective and can convey volumes more clearly than a plan view. These representations are constructed using three axes (X–Y–Z) in conjunction with a horizontal baseline, from which the parallels are projected at different angles.

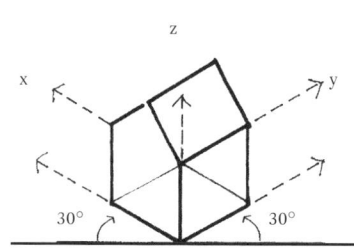
Baseline

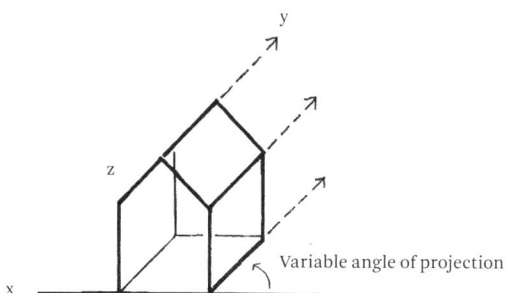
Variable angle of projection

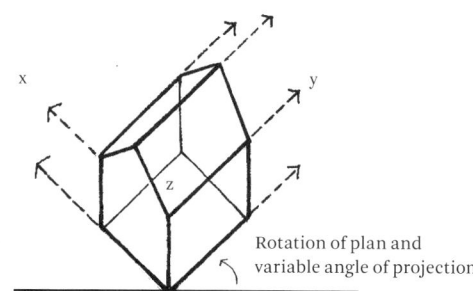
Rotation of plan and variable angle of projection

Parallel projections
Introduction
Isometric

Isometric

An isometric projection makes all three sides of an object visible, each shown with equal emphasis. Constructing an isometric projection is clearly defined: parallel lines are projected with a 30° angle to the right and left of the horizontal baseline. Drawing an isometric by hand requires all individual dimensions to be measured in plan and then projected accordingly. Direct projection from plan is not possible. The isometric has a very subtle and easily legible effect due to the relatively gentle angles of projection, offering the viewer an interesting view as if hovering above the entire composition.

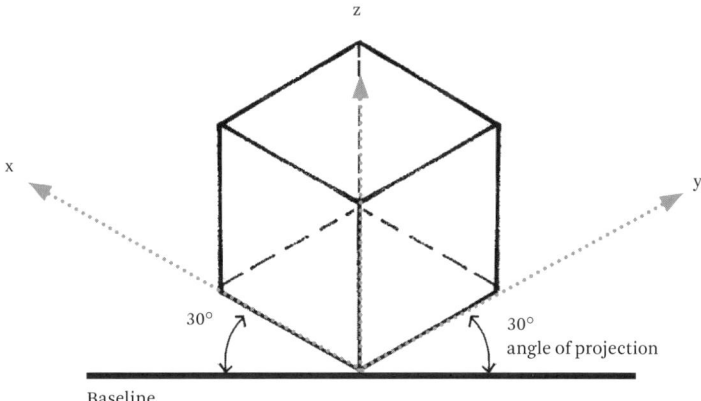

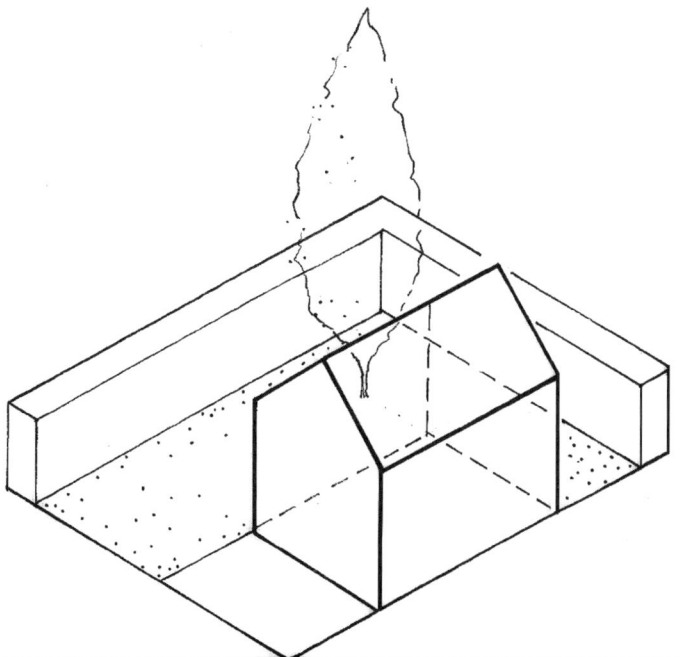

Parallel projections

Elevation oblique

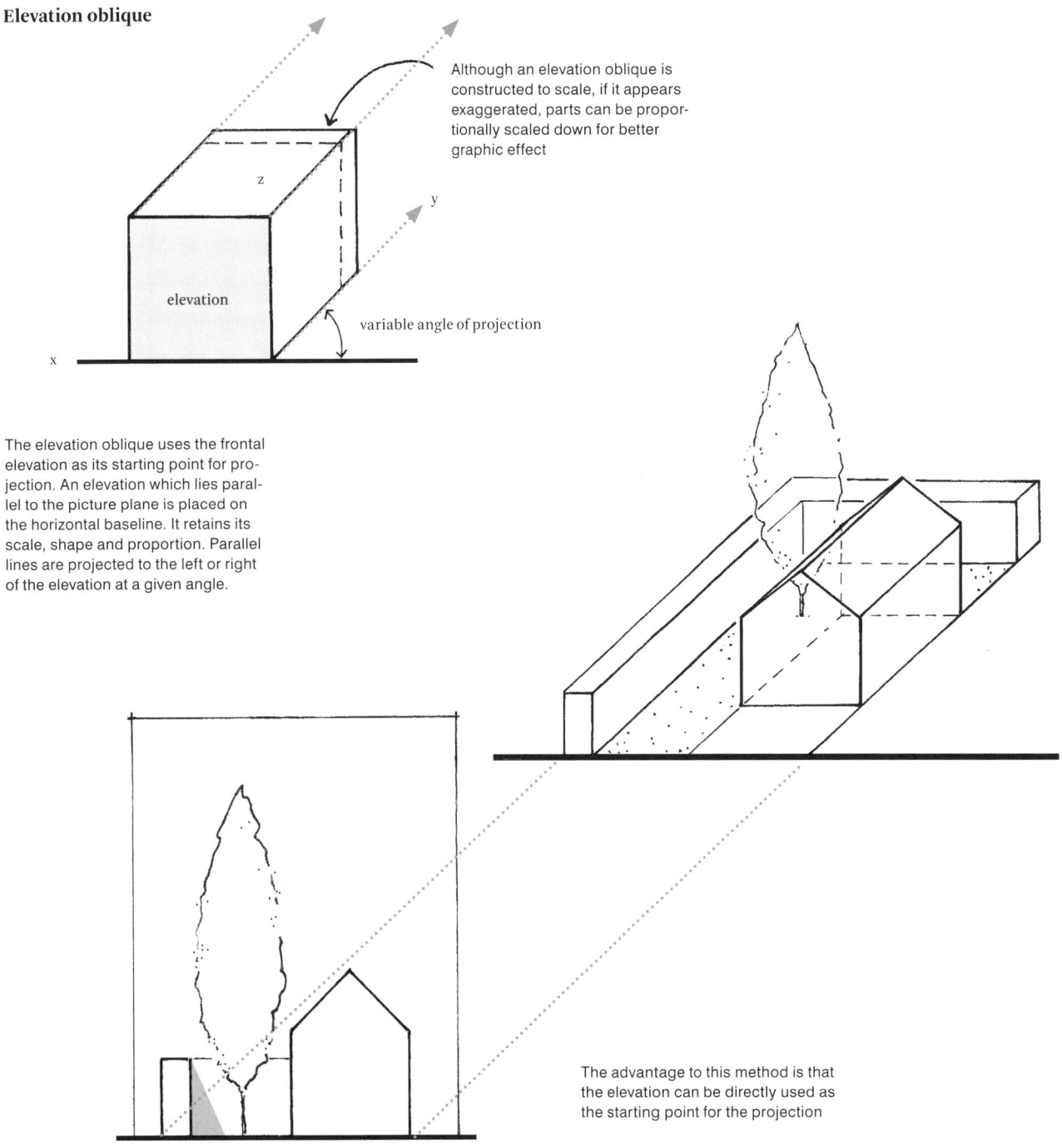

Although an elevation oblique is constructed to scale, if it appears exaggerated, parts can be proportionally scaled down for better graphic effect

variable angle of projection

The elevation oblique uses the frontal elevation as its starting point for projection. An elevation which lies parallel to the picture plane is placed on the horizontal baseline. It retains its scale, shape and proportion. Parallel lines are projected to the left or right of the elevation at a given angle.

Elevation

The advantage to this method is that the elevation can be directly used as the starting point for the projection

Elevation oblique
Introduction and construction

Axonometric projection
Introduction and construction
Circles
Trees
Vegetation
Construction steps

5

Axonometric projection / plan oblique

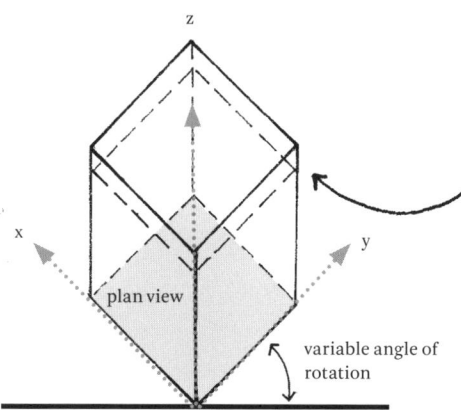

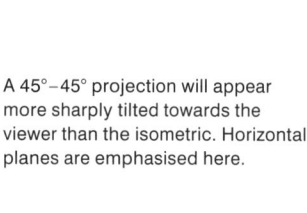

The axonometric projection is always constructed to scale, however it is acceptable to proportionally scale down vertical heights if they appear exaggerated.

The axonometric projection is easy to construct, as it projects directly from the scaled plan view. The plan is rotated in conjunction with a horizontal baseline, which offers a wide range of possible views and emphasis.

A 45°–45° projection will appear more sharply tilted towards the viewer than the isometric. Horizontal planes are emphasised here.

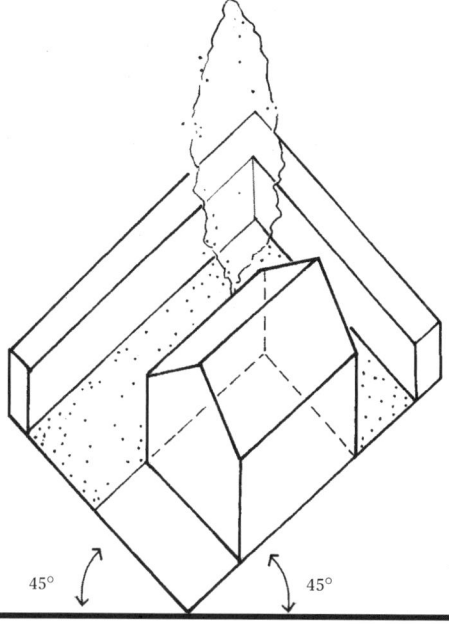

A 30°–60° projection will emphasise one of the vertical planes over the other. Axonometric projections are not bound to any given angles of projection, so a plan can be angled as required to show important content. Since overlaps occur in all parallel projections, it is important to make sure that nothing important is blocked in the final image.

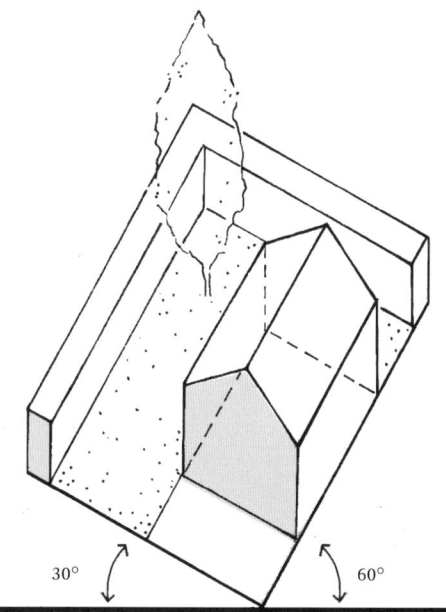

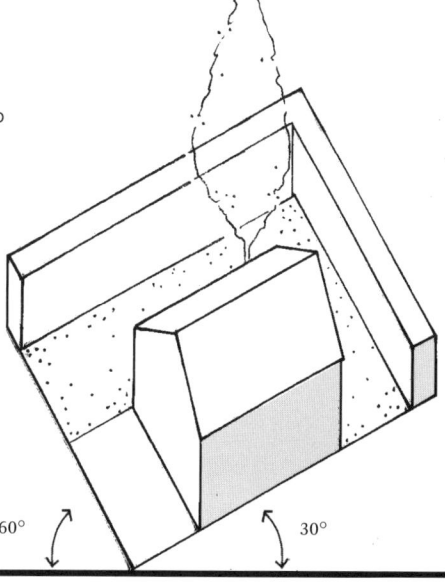

161

Parallel projections

Circles in axonometric projections

Circles on the horizontal plane will retain their true form. The same applies to the circles on the vertical plane parallel to the picture plane in the elevation oblique. Projecting circles to form a cylinder is easily done by simply projecting upwards (to a given height) a second circle of the same size. The two circles can then be connected using vertical lines, which form the outline of the cylinder.

Circles lying in the projected planes will form ellipses. To order to draw a circle in an axonometric it is advisable to circumscribe the circle using a square. The square can be further subdivided into quadrants using diagonals. These subdivisions can also be applied to the projected plane. The points where the circle intersects the different divisions guide the construction of the ellipse.

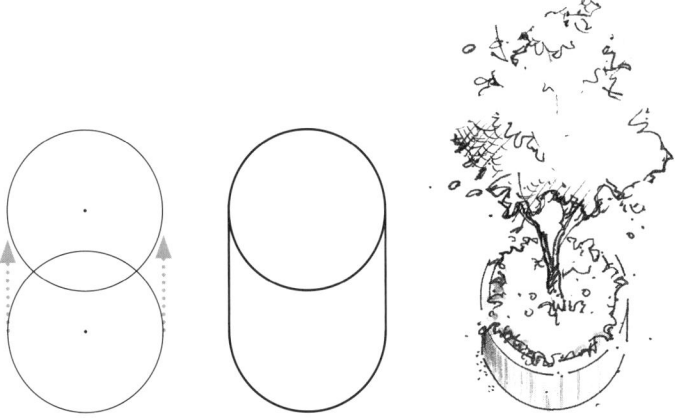

Constructing a cylinder in axonometric projection is simple. The circle on the horizontal ground plan retains its true form and is projected upwards.

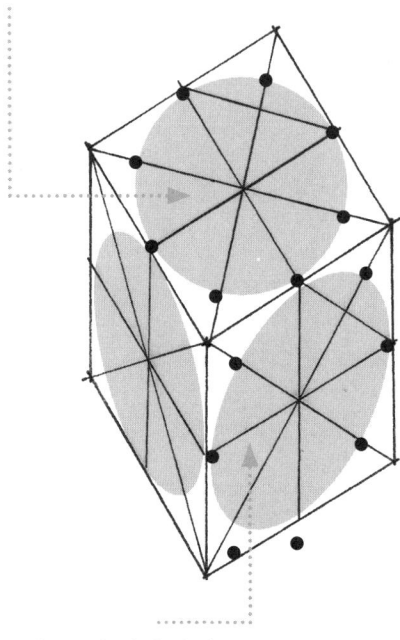

The circle in the upper horizontal plane remains a circle...

...whereas the circles in the projected, non-frontal planes appear as ellipses

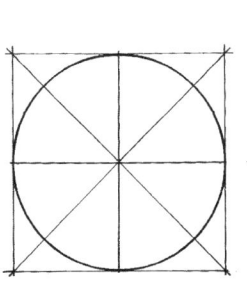

Intersection points can be used to help draw the circle

Axonometric projection
Introduction and construction
Circles
Trees
Vegetation
Construction steps

5

Trees in an axonometric projection
It is important to remember that, in axonometric projections, trees are viewed from above. They appear much more voluminous and three-dimensional than in an elevation, where they are flatly projected and viewed from one side.

Elevation

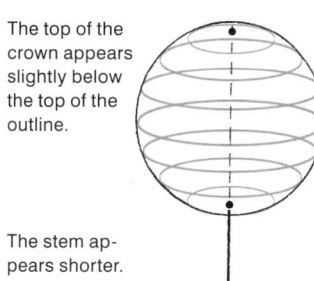

The top of the crown appears slightly below the top of the outline.

The stem appears shorter.

Axonometric projection

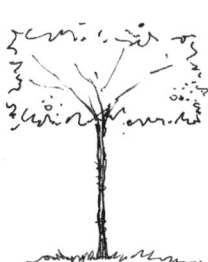

Elevation

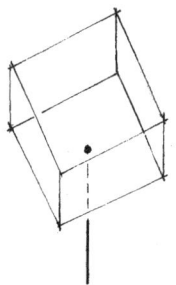

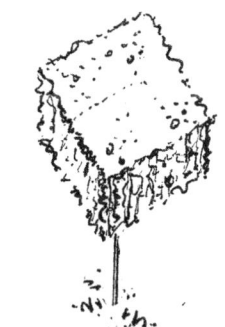

Axonometric

Elevation

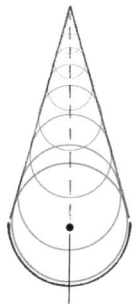

Axonometric

163

Parallel projections

Vegetation in axonometric projections
The method for drawing shrubs and smaller woody plants in axonometric projections is similar to that for drawing trees. The basic form in the plan view is projected upwards to a measured height. Their appearance is more voluminous than in elevation.

By adding a light source and graphically enhancing them with foliage textures, the forms are given a further convincing three-dimensional quality.

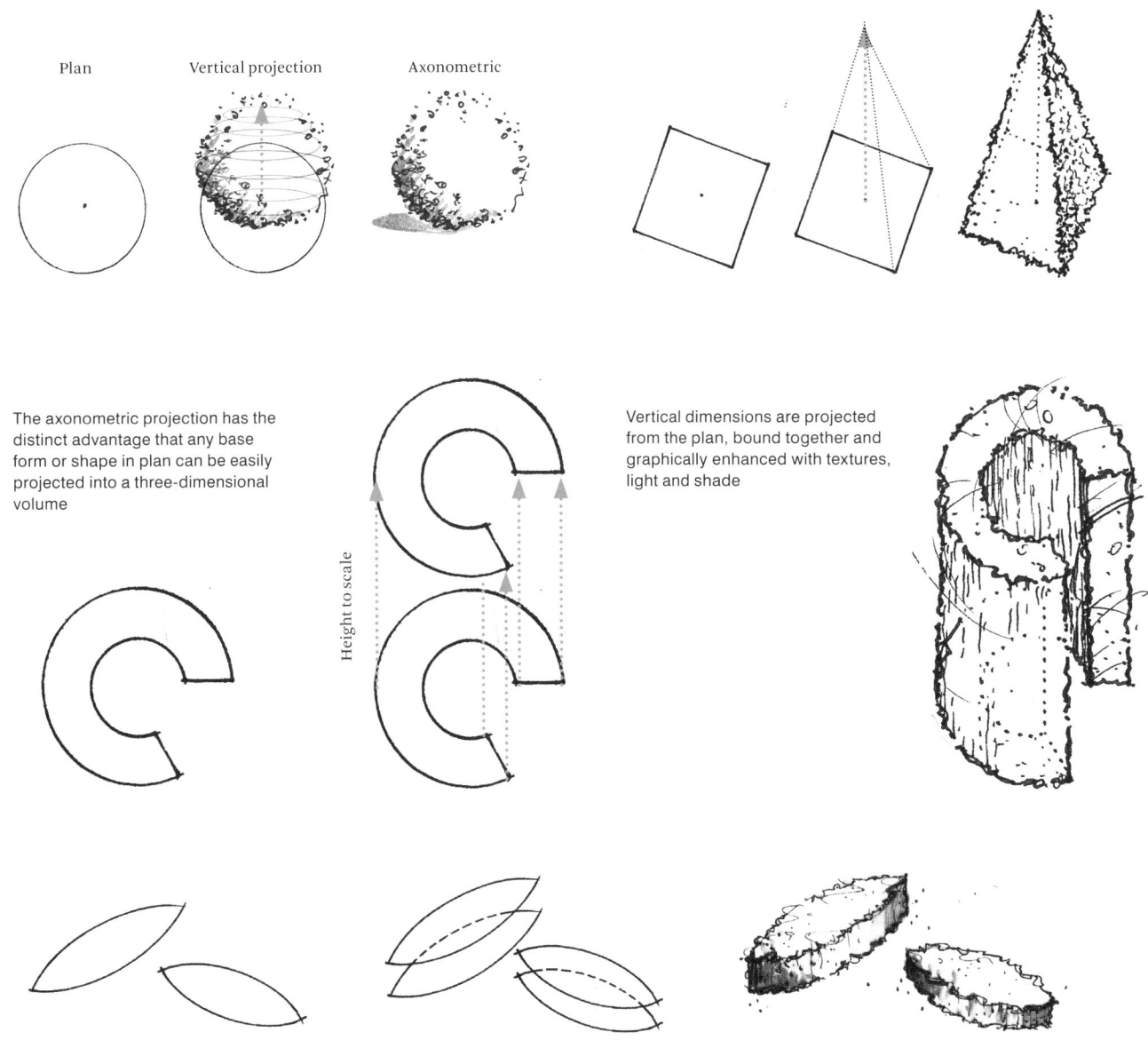

The axonometric projection has the distinct advantage that any base form or shape in plan can be easily projected into a three-dimensional volume

Vertical dimensions are projected from the plan, bound together and graphically enhanced with textures, light and shade

Axonometric projection
Introduction and construction
Circles
Trees
Vegetation
Construction steps

5

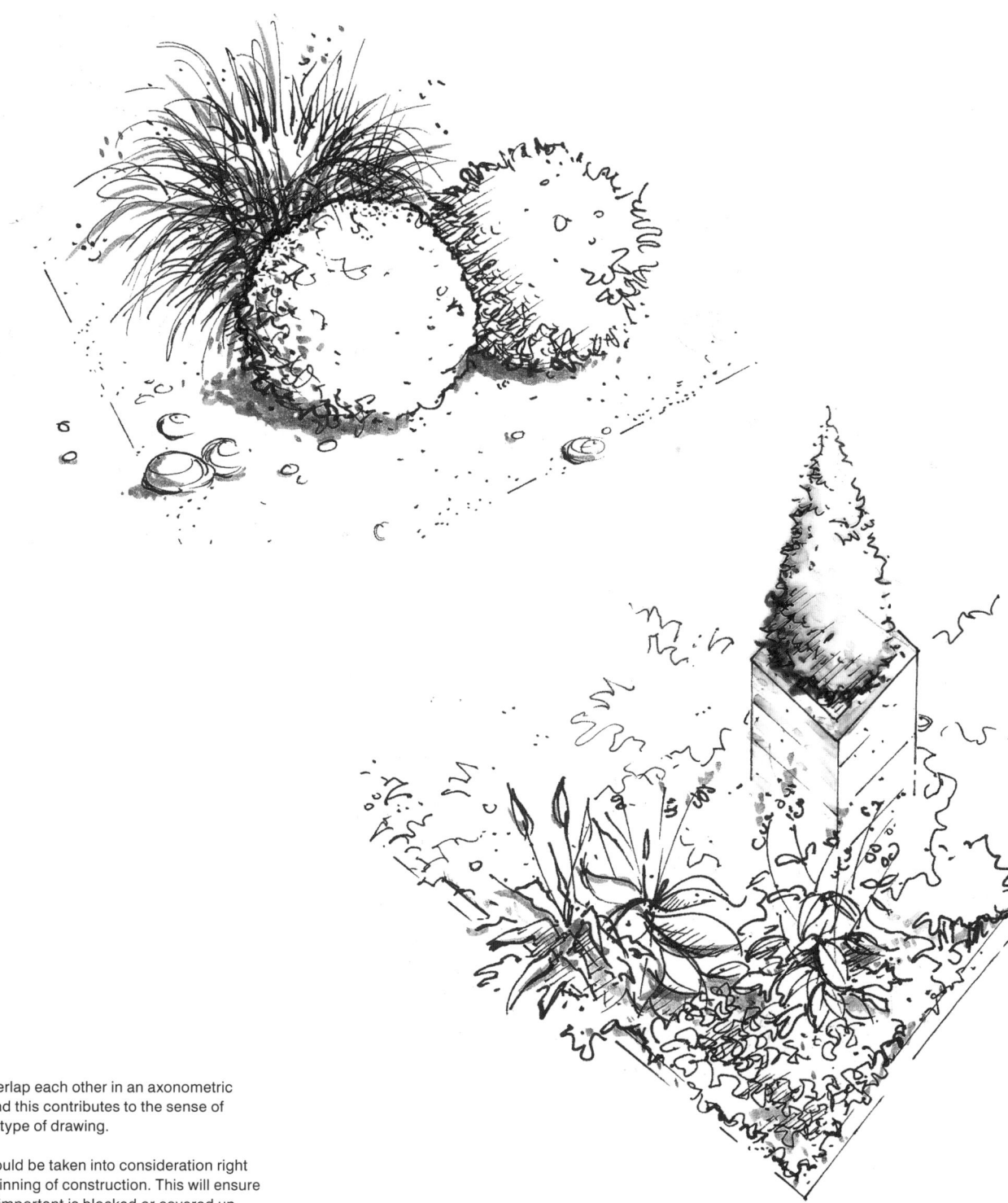

Elements overlap each other in an axonometric projection and this contributes to the sense of depth in this type of drawing.

Overlaps should be taken into consideration right from the beginning of construction. This will ensure that nothing important is blocked or covered up.

Parallel projections

Constructing an axonometric projection from plan: The steps
An axonometric projection begins with a plan and elevation to scale.

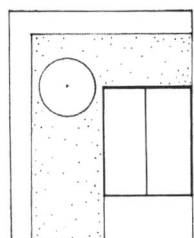

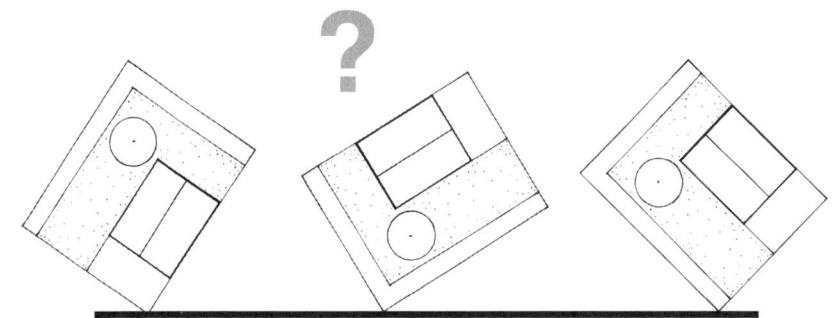

The plan is rotated in conjunction with a baseline, making sure that the angle of projection will allow important information to appear

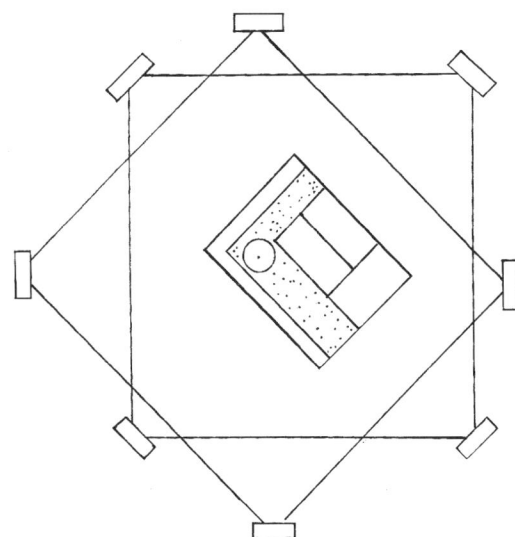

Once the angle of rotation is defined, a second sheet of either sketch paper or vellum is fastened over the plan

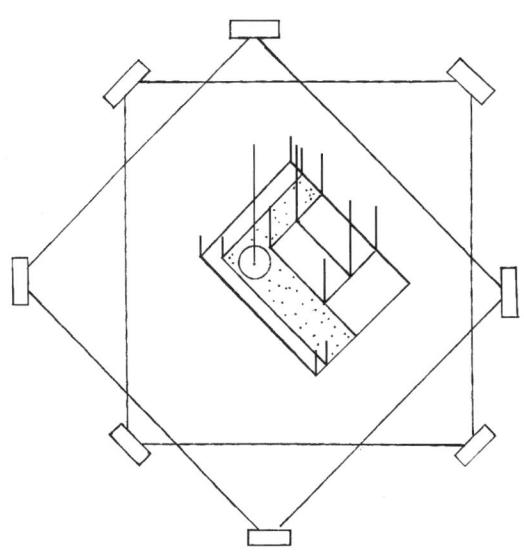

Object edges are projected vertically according to the heights indicated in the elevation

Axonometric projection
Introduction and construction
Circles
Trees
Vegetation
Construction steps

5

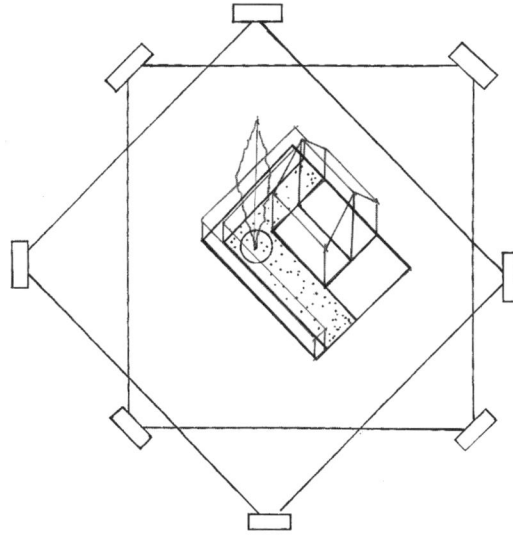

The lines are bound together to result in the three dimensional construction

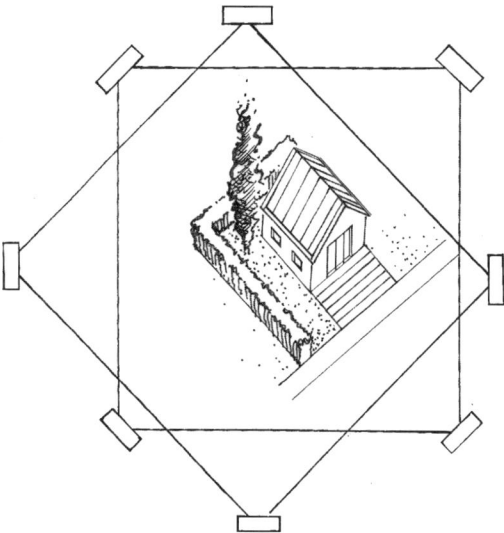

A further layer of paper is overlaid onto the construction sketch, which is then redrawn and enriched with details

By adding a light source and rendering shade and shadows, the scene gains additional three-dimensional qualities

Parallel projections

In order to fully utilise the three-dimensional qualities of an axonometric projection, remember to rotate the plan in such a way as to ensure that all important elements can be seen unobstructed in the final image. The resulting axonometric drawing should always be easily legible.

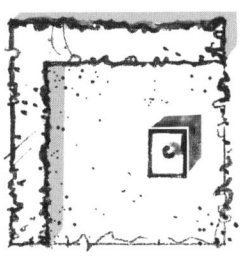 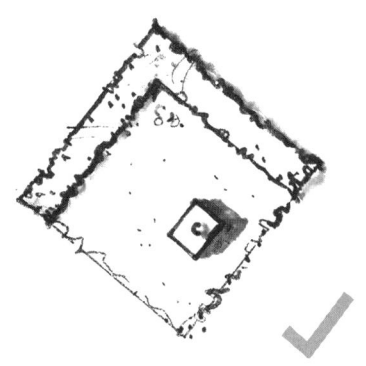 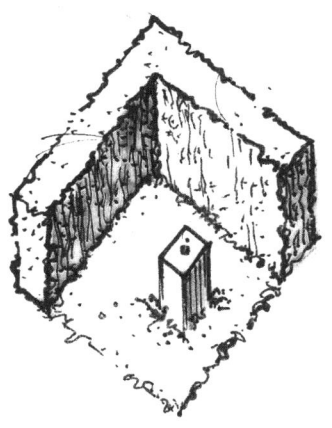

Not rotating the plan and simply projecting it vertically may appear strange to untrained eyes, since the axonometric does not seem to have three dimensions

In this example, the plan is rotated in such a way that information is lost. The rectangular form is obstructed by the hedge. The space is not legible.

Projecting the orthogonal plan sideways also appears strange and requires viewers to tilt their heads in order to read the space

 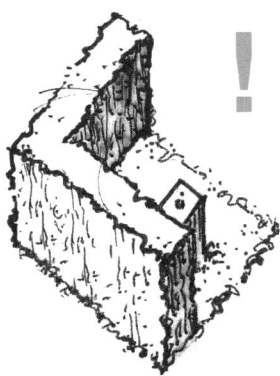 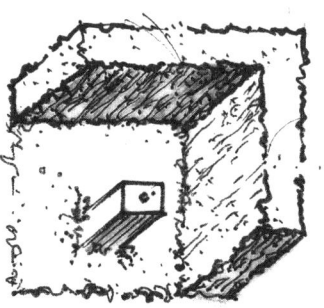

Axonometric projection
Introduction and construction
Circles
Trees
Vegetation
Construction steps

5

Although these days the construction of complex axonometric models are frequently handed over to CAD programmes, the sketch method is still a simple and quick way to communicate a design's features in three dimensions. Axonometric projections can be constructed at all scales, from construction details to large scale public spaces. They have a distinct advantage over the plan and elevation, since they are easily understood by the viewers, both trained professionals and laypersons alike.

6

Perspective

Perspective projection
172 **Introduction**
174 **Characteristics**
178 **Vanishing points**
180 **Types of perspective**
182 **Coordinates and sightlines**
183 **Cone of vision**
184 **Constructing a perspective grid**
185 **Diagonals**
186 **Horizon line and pictorial effect**
188 **Stairs and ramps**
189 **Reflections**
190 **Repetitive forms and dimensions**
192 **Circles**
193 **Simple shadows**

Construction methods
194 **From the plan view**
198 **Using a perspective grid**
204 **From photos**
206 **Drawing freehand perspectives**
208 **Estimating proportions**
210 **Freehand one-point perspective**
212 **Freehand two-point perspectives**
214 **Atmospheric perspective**
216 **Graphic emphasis**

Perspective

Drawing perspectives: Seeing and understanding space
In order to draw perspectives correctly, it is important to understand how we see space. In order to master the perspective, knowing its principles and characteristics is absolutely essential. This is especially so for landscape architects who concern themselves with designing and communicating space rather than objects or buildings. Perspectives follow principles which are rooted both in descriptive geometry and in painting.

These principles are valid when drawing perspectives at a desk in studio and when drawing outdoors, on site. Perspective projections offer a sensory view of a space, as opposed to the more mechanical views found in paraline or orthographic projections. If a perspective space is not constructed correctly, it will be impossible to rescue it even with most sophisticated graphic rendering.

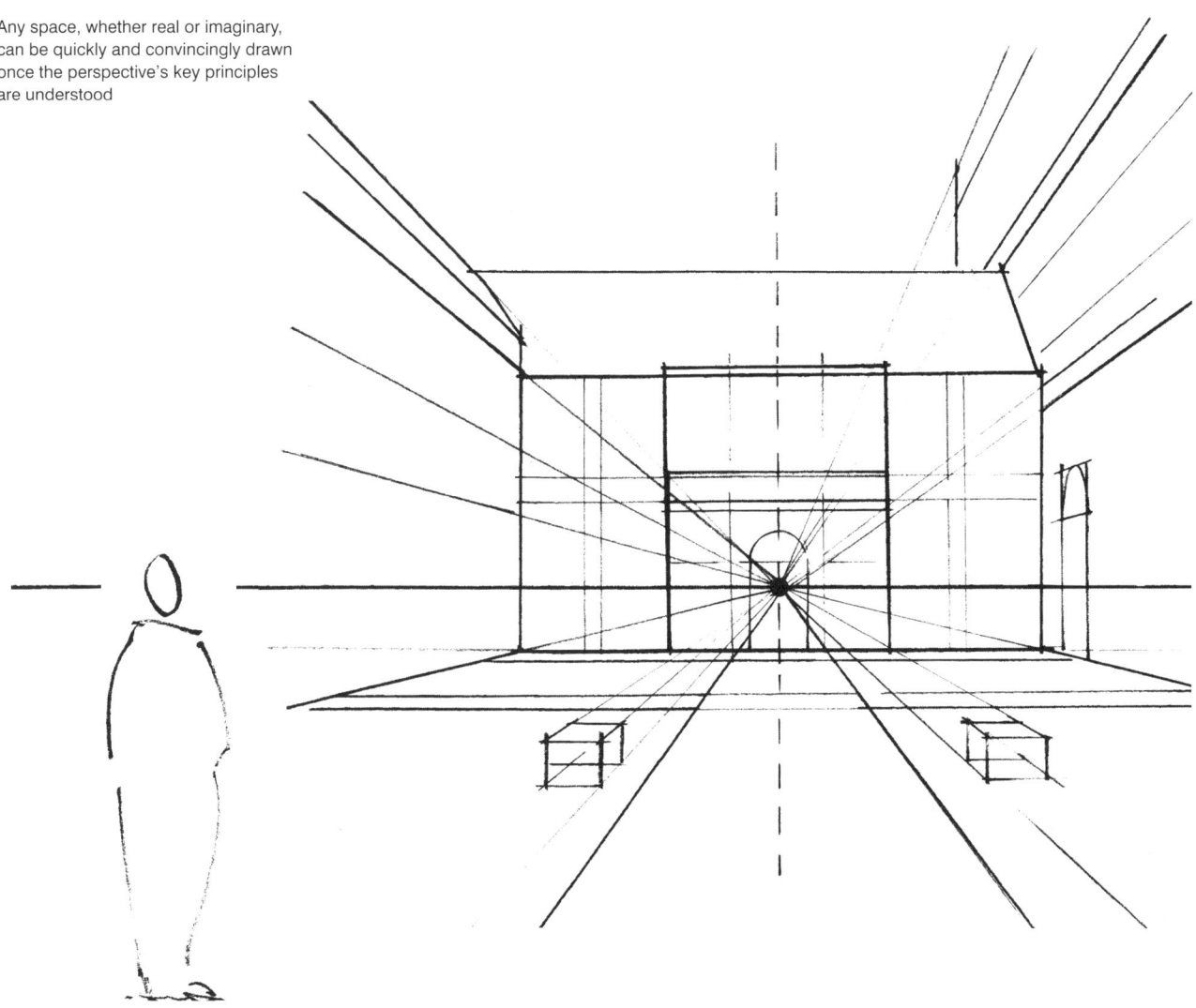

Any space, whether real or imaginary, can be quickly and convincingly drawn once the perspective's key principles are understood

6

Axonometric projection
Introduction
Characteristics
Vanishing points
Types of perspective
Coordinates and sightlines
Cone of vision
Constructing a perspective grid
Diagonals
Horizon line and pictorial effect
Stairs and ramps
Reflections
Repetitive forms and dimensions
Circles
Simple shadows

Two main types of perspective will be discussed here, as they appear most frequently in the design and planning disciplines: linear perspective and atmospheric perspective.

The topic of perspective is very important, and is the subject of countless books. This chapter will only briefly show the different types of perspective, how they are used in landscape architecture and the basic methods for their construction.

Perspective

Linear perspective: Basic principles

Although the linear perspective was well known in Antiquity, it became less important in the visual arts during medieval times. It wasn't until its rediscovery Renaissance, that it was extensively developed as an art and as a science. Perspective knowledge allowed countless masterpieces in art and architecture to happen and is still very relevant in design development and presentation today.

Linear perspective follows clear and defined rules and principles. As can be deduced from the name, linear perspective deals primarily with lines and how they behave in conjunction with these principles. It is therefore extremely useful for constructing and presenting three dimensional built structures and spaces, as they would be naturally seen and experienced by a viewer.

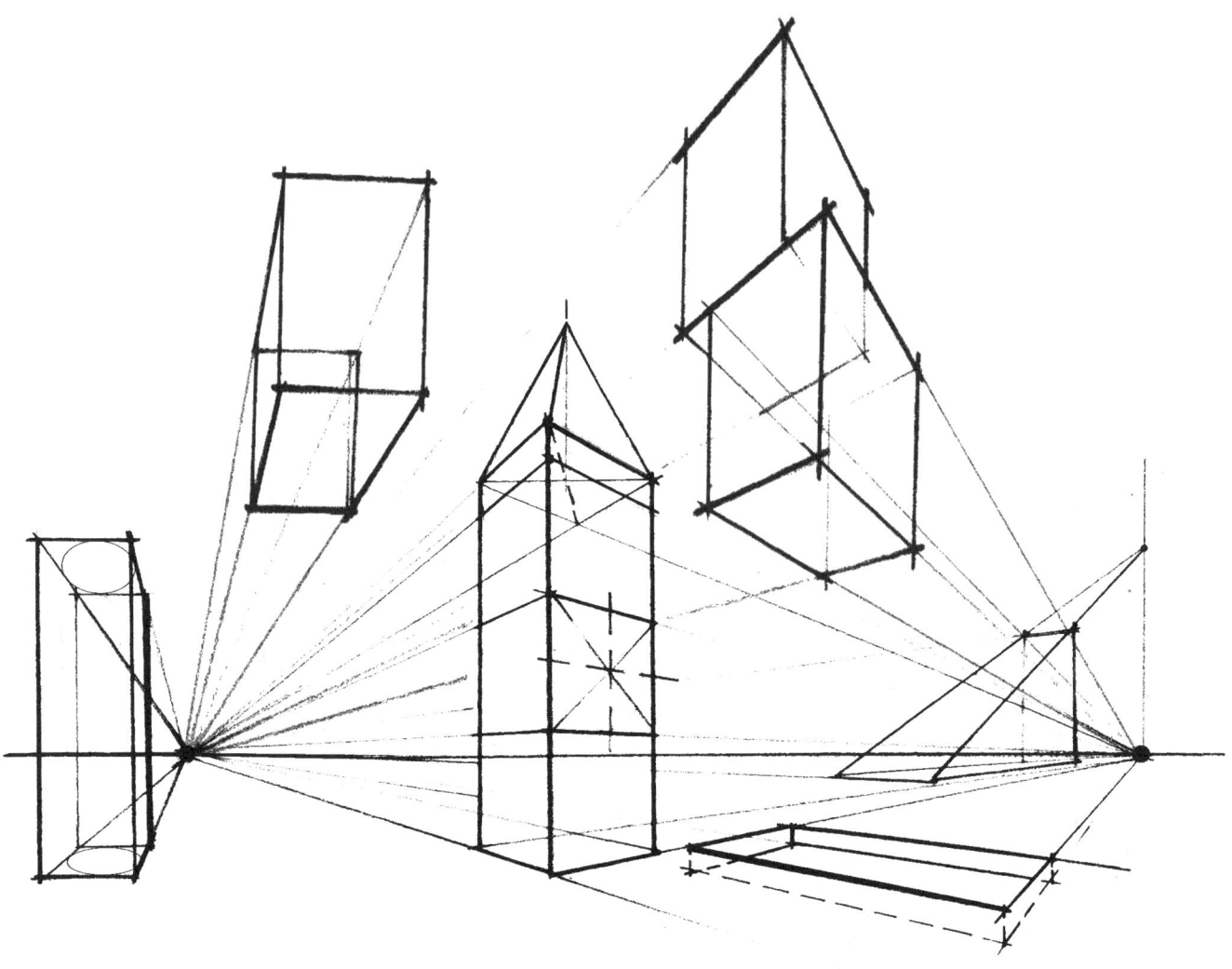

The convergence of parallel lines as they recede from the viewer is one of the most common characteristics of a linear perspective

Axonometric projection
Introduction
Characteristics
Vanishing points
Types of perspective
Coordinates and sightlines
Cone of vision
Constructing a perspective grid
Diagonals
Horizon line and pictorial effect
Stairs and ramps
Reflections
Repetitive forms and dimensions
Circles
Simple shadows

6

A linear perspective assumes that a viewer is standing at a fixed position in space and looking at a scene directly in front of him. This is often referred to as monocular vision. In contrast to reality, the viewer sees the scene as if through a single eye and thus has central sightline with a limited cone of vision. This cone has a radius of approximately 60°–90°. Another key assumption in perspective is that a picture plane, like an upright transparent canvas or frame, is located between the viewer and the space he is looking straight at. This picture plane is essentially our own canvas or paper block upon which the perspective scene is drawn.

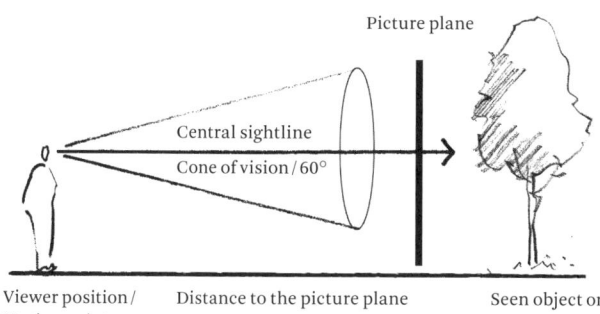

Every linear perspective possesses its own horizon line, equal to the height of the viewer's eye level from his station point on the ground plane. The horizon line is not normally visible, unless we are looking at the ocean or are located high up on a mountain. It does play an important role in how perspective is perceived and should be carefully considered when constructing perspectives.

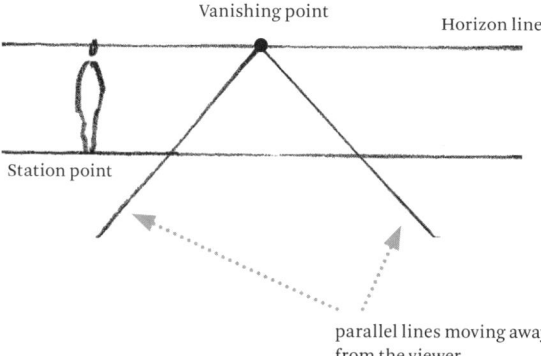

Unlike paraline views, not all parallel lines remain parallel. Lines moving away from the viewer appear to converge at specific points on the horizon line. These are called vanishing points and occur where the viewer's sightline meets the horizon line on the picture plane. Most of us will have seen a picture of railway tracks which seem to extend into infinity towards a horizon line far in the distance. This shows how parallel lines can behave in linear perspective and distinguishes perspective from orthographic and parallel projections.

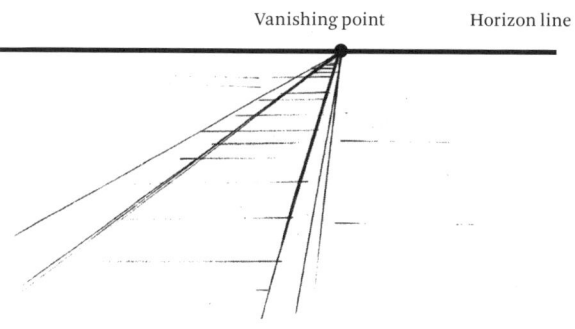

Perspective

Linear perspective: Principles and characteristics

The methodology for constructing perspectives follow distinct rules of geometry and are based on the concept of 'seeing through' an imaginary, transparent picture plane. With the help of the viewer's sightlines, all elements behind this plane are projected onto the plane to form the resulting perspective image. The diagram below shows the set up, which must be kept in mind when constructing a perspective.

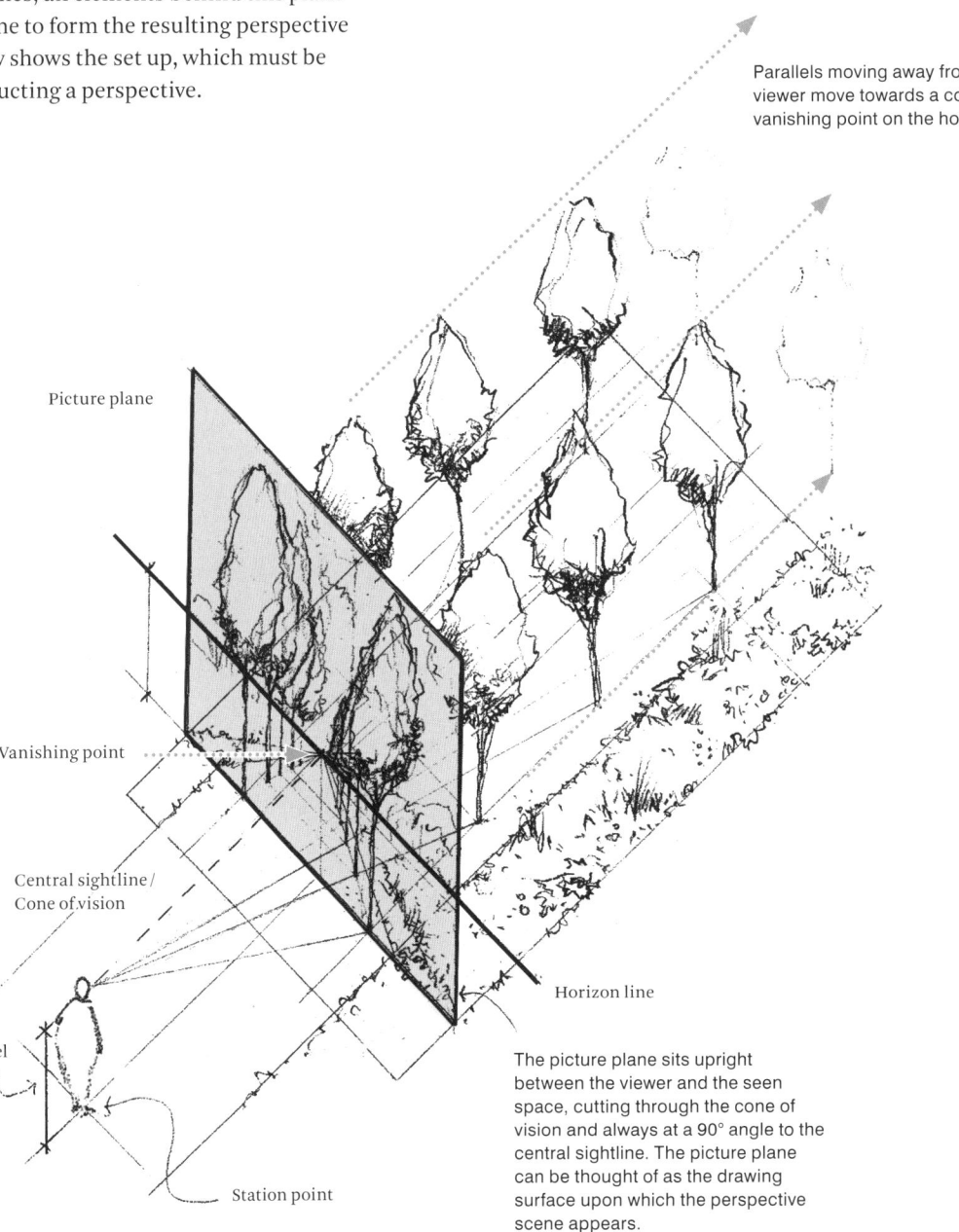

Parallels moving away from the viewer move towards a common vanishing point on the horizon line

Picture plane

Vanishing point

Central sightline / Cone of vision

Viewer with a fixed position and eye level

Station point

Horizon line

The picture plane sits upright between the viewer and the seen space, cutting through the cone of vision and always at a 90° angle to the central sightline. The picture plane can be thought of as the drawing surface upon which the perspective scene appears.

Axonometric projection
Introduction
Characteristics
Vanishing points
Types of perspective
Coordinates and sightlines
Cone of vision
Constructing a perspective grid

Diagonals
Horizon line and pictorial effect
Stairs and ramps
Reflections
Repetitive forms and dimensions
Circles
Simple shadows

6

Linear perspectives have clearly visible characteristics.

They all have converging lines which seem to move towards a common vanishing point (or points) on the distant horizon.

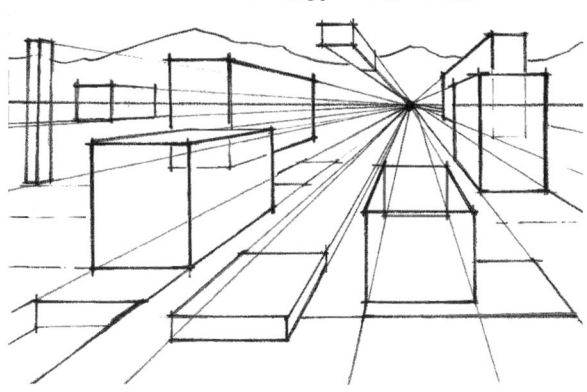

Vanishing point in the distance

A perspective scene often has overlapping objects and foreshortened planes.

Objects of the same size appear to get smaller as they get closer to the vanishing point. Vertical lines remain vertical; however they too appear to decrease in size in the distance. These principles add pictorial effects and a distinctive atmospheric depth to a perspective scene.

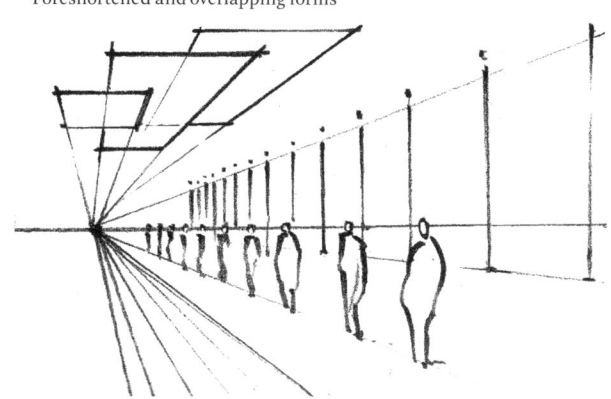

Foreshortened and overlapping forms

Converging lines

Same-sized objects decrease in size as they get closer to the vanishing point

These characteristics all come together to form a sort of optical illusion. Space appears to extend well beyond the two-dimensional picture frame. Illustrating the scene with a light source and shadows also increases this perceived depth and spatial effect. The paper surface seems to disappear, resulting in a convincing three dimensional illusion which closely corresponds to our experience of space.

Perspective

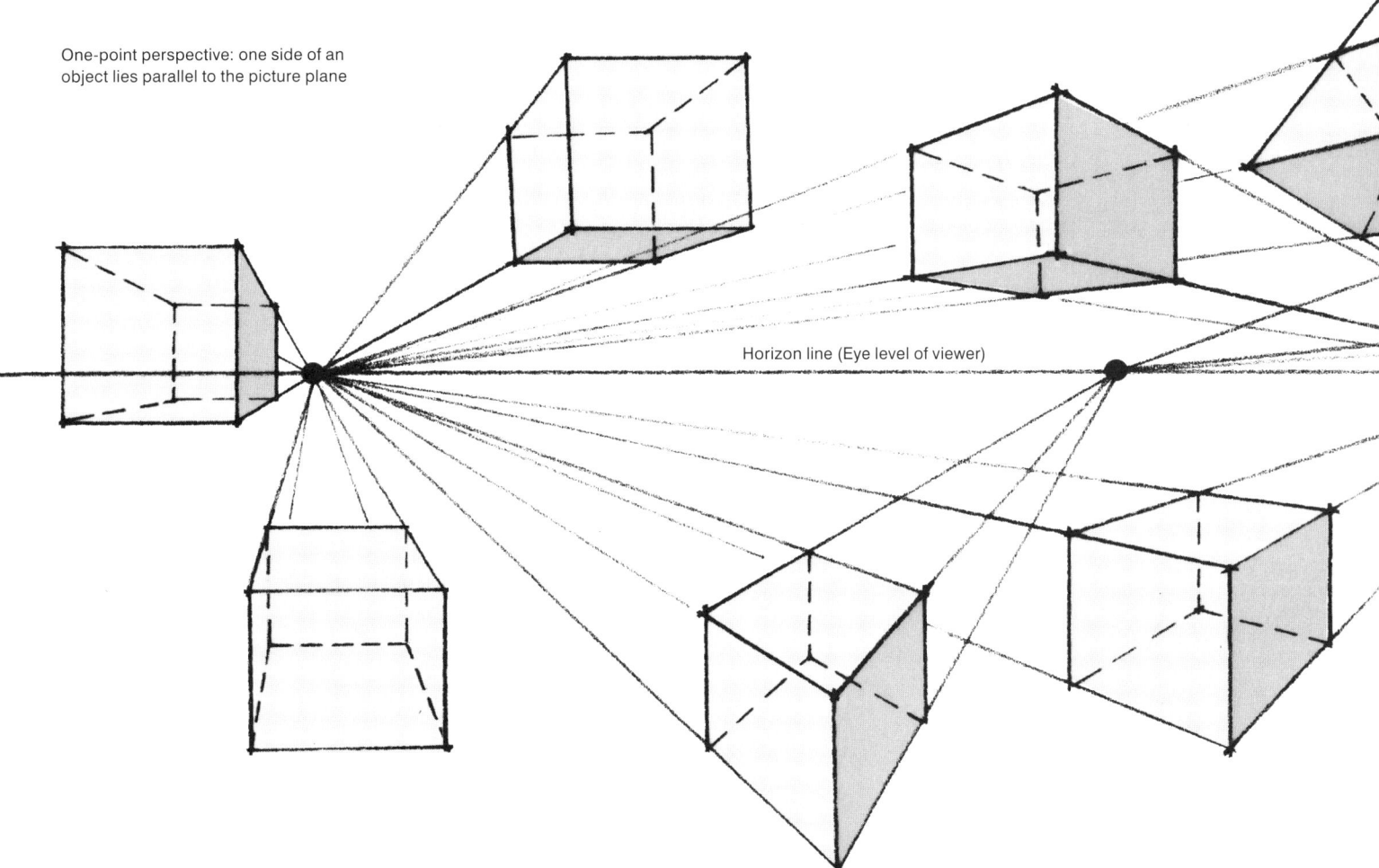

One-point perspective: one side of an object lies parallel to the picture plane

Horizon line (Eye level of viewer)

Two-point perspective: no side of the object is parallel to the picture plane

Axonometric projection
Introduction
Characteristics
Vanishing points
Types of perspective
Coordinates and sightlines
Cone of vision
Constructing a perspective grid
Diagonals
Horizon line and pictorial effect
Stairs and ramps
Reflections
Repetitive forms and dimensions
Circles
Simple shadows

6

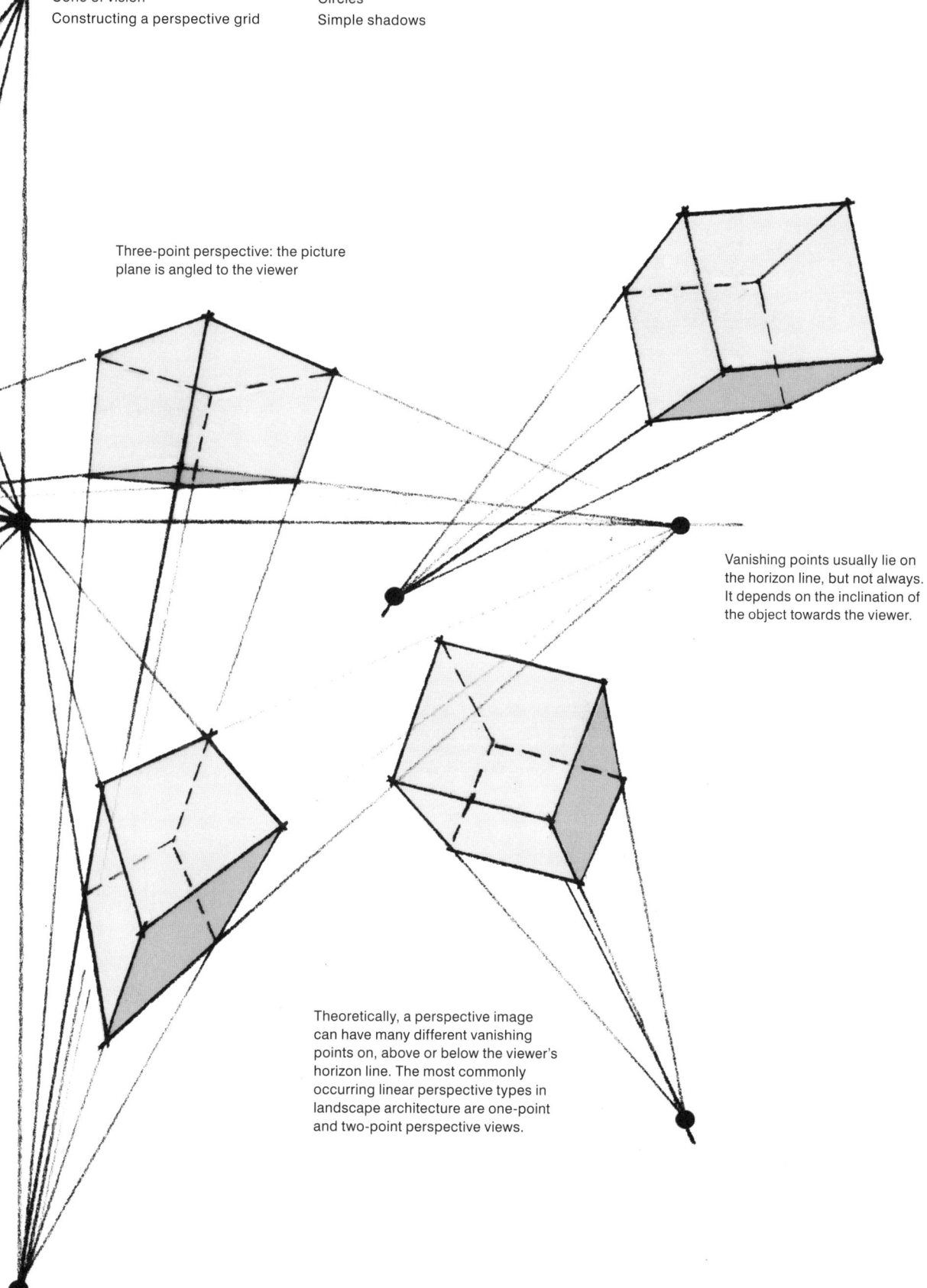

Three-point perspective: the picture plane is angled to the viewer

Vanishing points usually lie on the horizon line, but not always. It depends on the inclination of the object towards the viewer.

Theoretically, a perspective image can have many different vanishing points on, above or below the viewer's horizon line. The most commonly occurring linear perspective types in landscape architecture are one-point and two-point perspective views.

Perspective

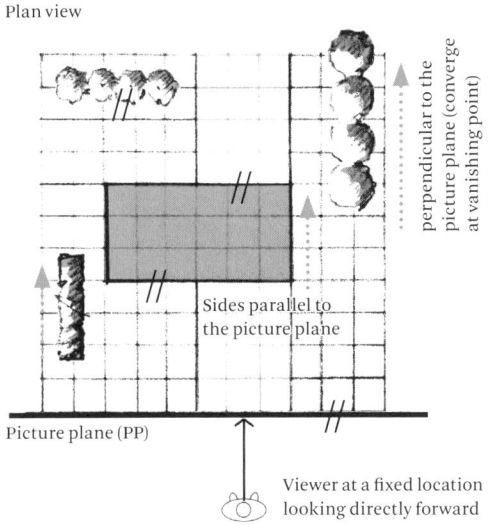

Plan view

Picture plane (PP)

Viewer at a fixed location looking directly forward

Sides parallel to the picture plane

perpendicular to the picture plane (converge at vanishing point)

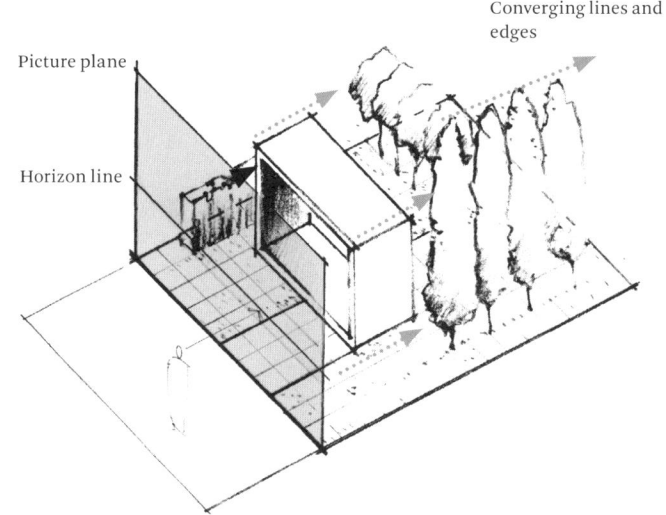

Picture plane

Horizon line

Converging lines and edges

In a one-point perspective, all parallel lines and planes which are parallel to the picture plane (PP) remain parallel. They will retain their proportional forms but reduce in size as they get closer to the vanishing point (VP).

This applies to the vertical elements, such as the edges of the built feature, the trimmed hedge and the line of poplar trees. They too diminish in size as they get closer to the vanishing point and appear further away from the viewer.

The resulting perspective scene clearly shows how elements which are not parallel to the picture plane behave. These lines, in this case the edges of the pathway, the upper and lower side edges of the built feature and hedge, as well as the line of trees, all appear to converge towards the central vanishing point

VP

Horizon line

Axonometric projection
Introduction
Characteristics
Vanishing points
Types of perspective
Coordinates and sightlines
Cone of vision
Constructing a perspective grid
Diagonals
Horizon line and pictorial effect
Stairs and ramps
Reflections
Repetitive forms and dimensions
Circles
Simple shadows

6

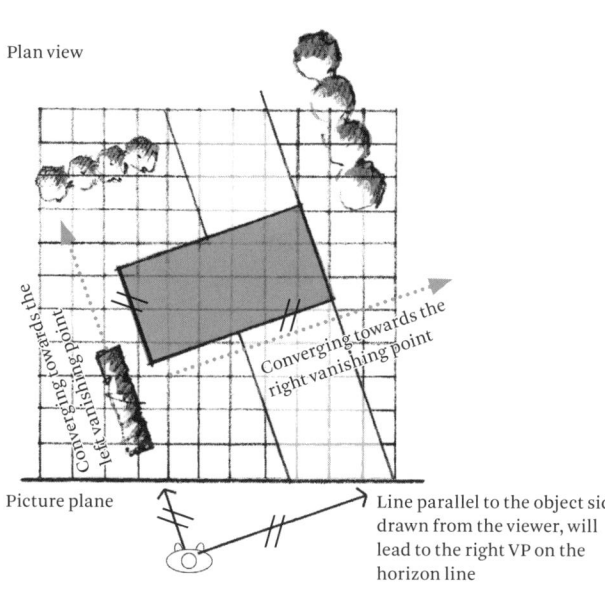

Plan view

Picture plane

Line parallel to the object side drawn from the viewer, will lead to the right VP on the horizon line

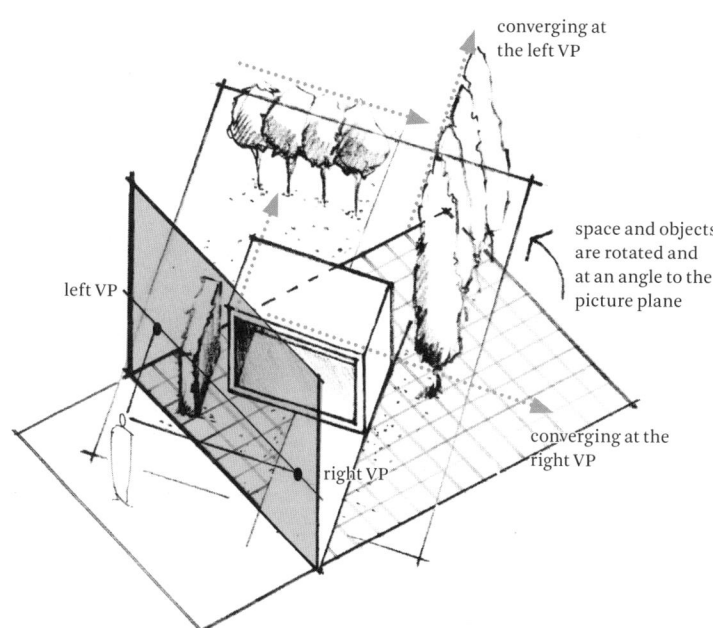

converging at the left VP

space and objects are rotated and at an angle to the picture plane

left VP

right VP

converging at the right VP

Two-point perspectives are more widely used. In a two-point perspective, the object of space is rotated and no plane is parallel to the picture plane. Only vertical edges are still parallel, everything else converges either to the left or to the right of the viewer.

The viewer remains at a fixed location and is looking straight ahead. The two vanishing points lie on the horizon line and can be found in conjunction with their angle of rotation. The vanishing points can be easily located by simply projecting two lines (parallel to the object's sides) from the viewer's station point in plan onto the picture plane.

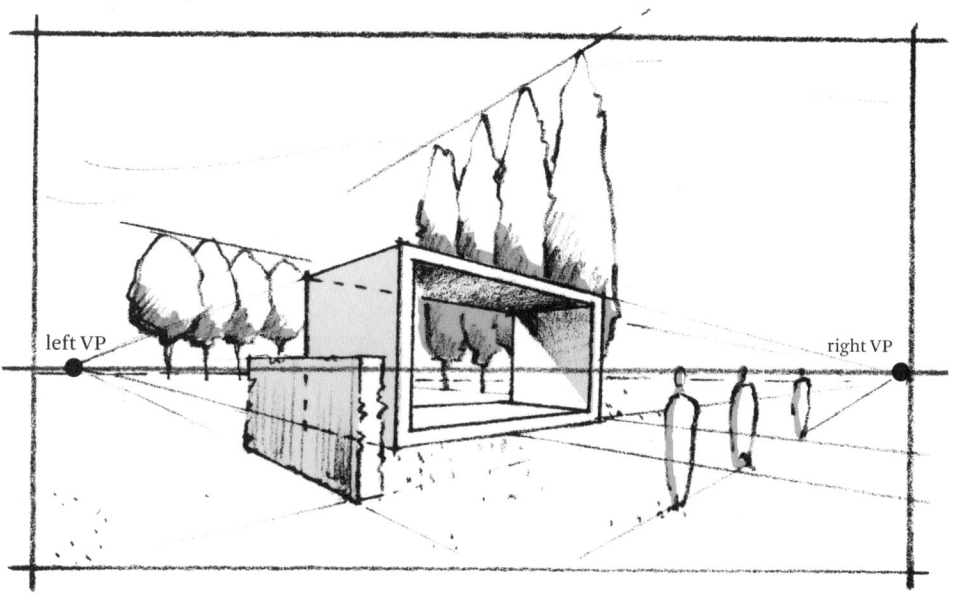

left VP

right VP

The resulting perspective scene demonstrates how the two sides of the built feature are rotated to the viewer, each converging at their own vanishing points. The scene appears less static and seems to possess a more natural view than the one-point perspective.

Perspective

Dimensioned space and depth illusion

A perspective view allows the viewer to look through the picture plane into a seemingly infinite space beyond

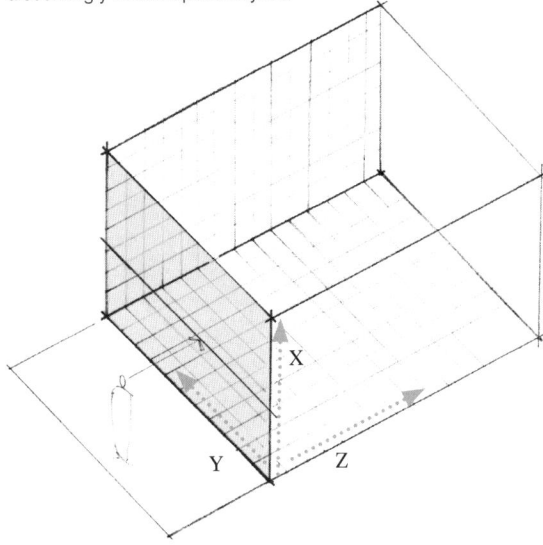

This spatial illusion is a three-dimensional coordinate system. Much like the parallel projections, it has three axes: the X, Y und Z axes. They subdivide the space into units and dimensions which will help to locate objects and planes within it. Together with the vanishing point, this dimensioned space appears to go on into infinity.

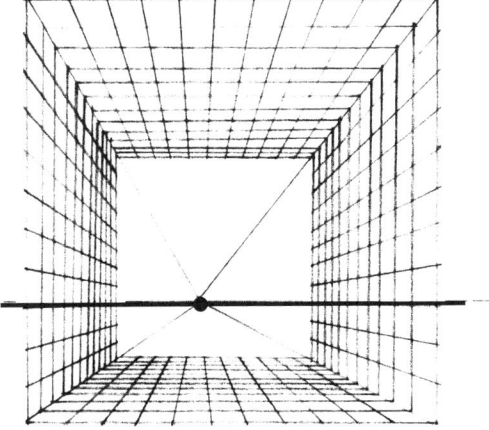

The viewer can locate objects in the space and uses sightlines to do so

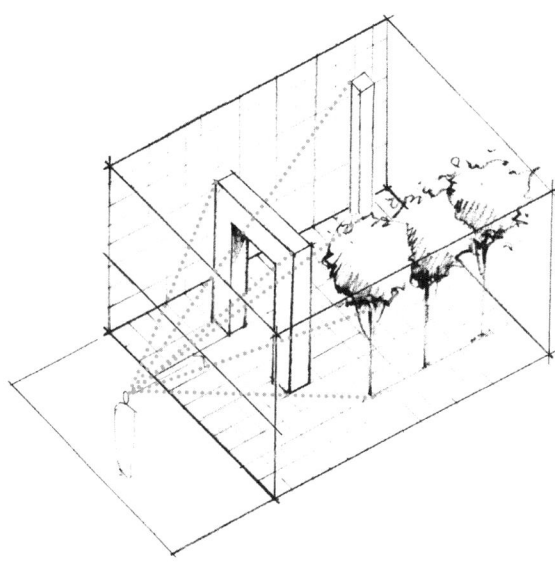

The resulting three-dimensional space is known as a perspective grid. This constructed space allows for quick production of spatial scenes, even without the help of a plan, section or elevation. A perspective grid is a handy tool when sketching and testing out spatial design variations.

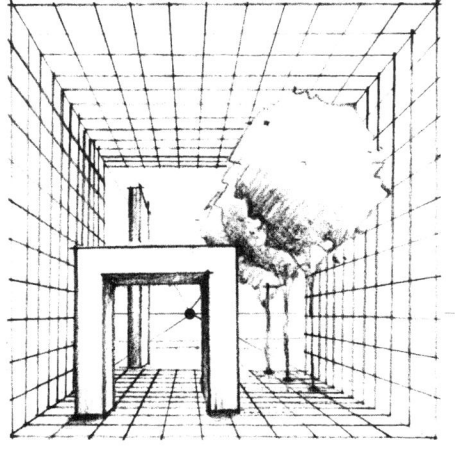

Axonometric projection
Introduction
Characteristics
Vanishing points
Types of perspective
Coordinates and sightlines
Cone of vision
Constructing a perspective grid

Diagonals
Horizon line and pictorial effect
Stairs and ramps
Reflections
Repetitive forms and dimensions
Circles
Simple shadows

Cone of vision

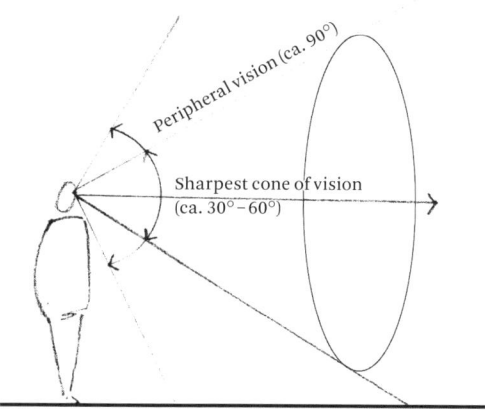

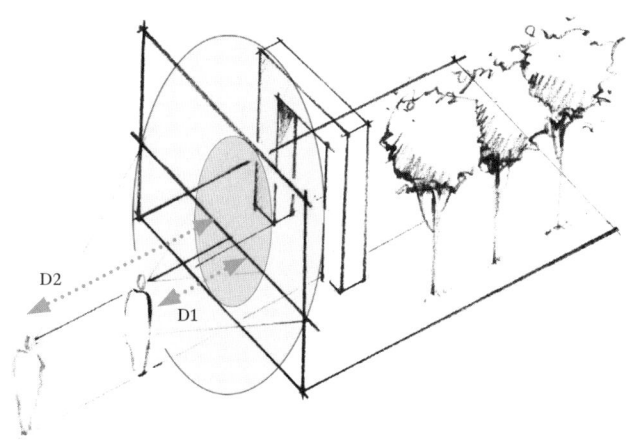

As briefly mentioned in the introduction, the size and sharpness of objects in a perspective drawing depend on distance between the viewer and the picture plane. To ensure that all important elements of a design scheme are clearly visible in the scene, the centre of interest should be placed within the cone of vision. This cone is formed by the sightlines radiating outwards from the central sightline, and can be assumed to be between 30° and 60°. In reality, our field of vision is much more extensive, with greater angles of peripheral vision, both vertically and horizontally. However, since the linear perspective is drawn using a single view point and one clear sightline, it cannot communicate a scene the way we would actually experience it. We are able to move our heads and our eyes much more dynamically.

The distance (D) between the viewer and the picture plane determines the cone of vision. The further away he stands, the larger it becomes. Important aspects of a scene or space should lie within the middle area of the cone. Placing objects too close to the viewer may not always be advisable because, as the cone is smaller here, only a small part of the foreground will be visible. More emphasis can be placed on the middleground and background, where the cone is considerably wider.

When constructing a perspective, the cone of vision can help determine what will be shown and what may be left out of a perspective drawing.

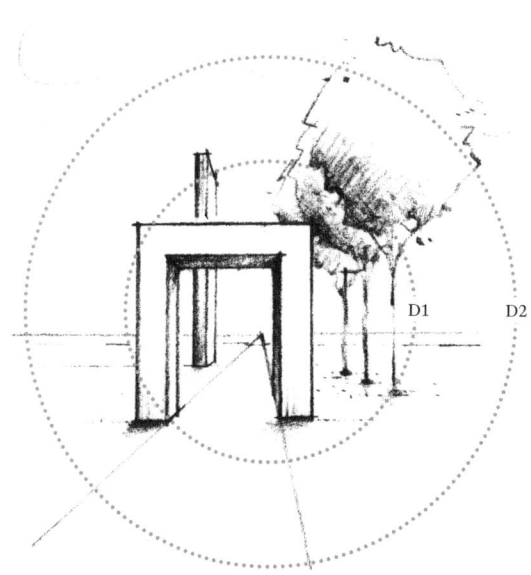

Perspective

Constructing a perspective grid

Constructing a perspective grid with one vanishing point is easy, since the picture plane can be used as a scale reference to measure all vertical and horizontal units. The increments along these lines, for example in metres or feet, are to scale. The picture plane forms a kind of frame in which the eye level and horizon line are drawn and the vanishing point (VP) is placed. All parallel lines perpendicular to the picture plane will converge at this one vanishing point. Drawing lines from the edges of the frame will begin to define the edges of the space, which can be easily extended if necessary. This diagonal point is nothing other than the vanishing point for the 45° lines which lie within square units. These lines enable us to scale the measurable depth units in perspective. Horizontal lines can be drawn where the diagonal lines intersect the converging lines. These are the edges of the square units in perspective. There is now grid of 1 m x 1 m squares on the ground plane which can be used to locate objects and boundaries within the space. Although each unit is exactly 1 m x 1 m, the squares in front of the picture plane appear larger than the squares behind it. These are smaller and more foreshortened as they recede towards the horizon line.

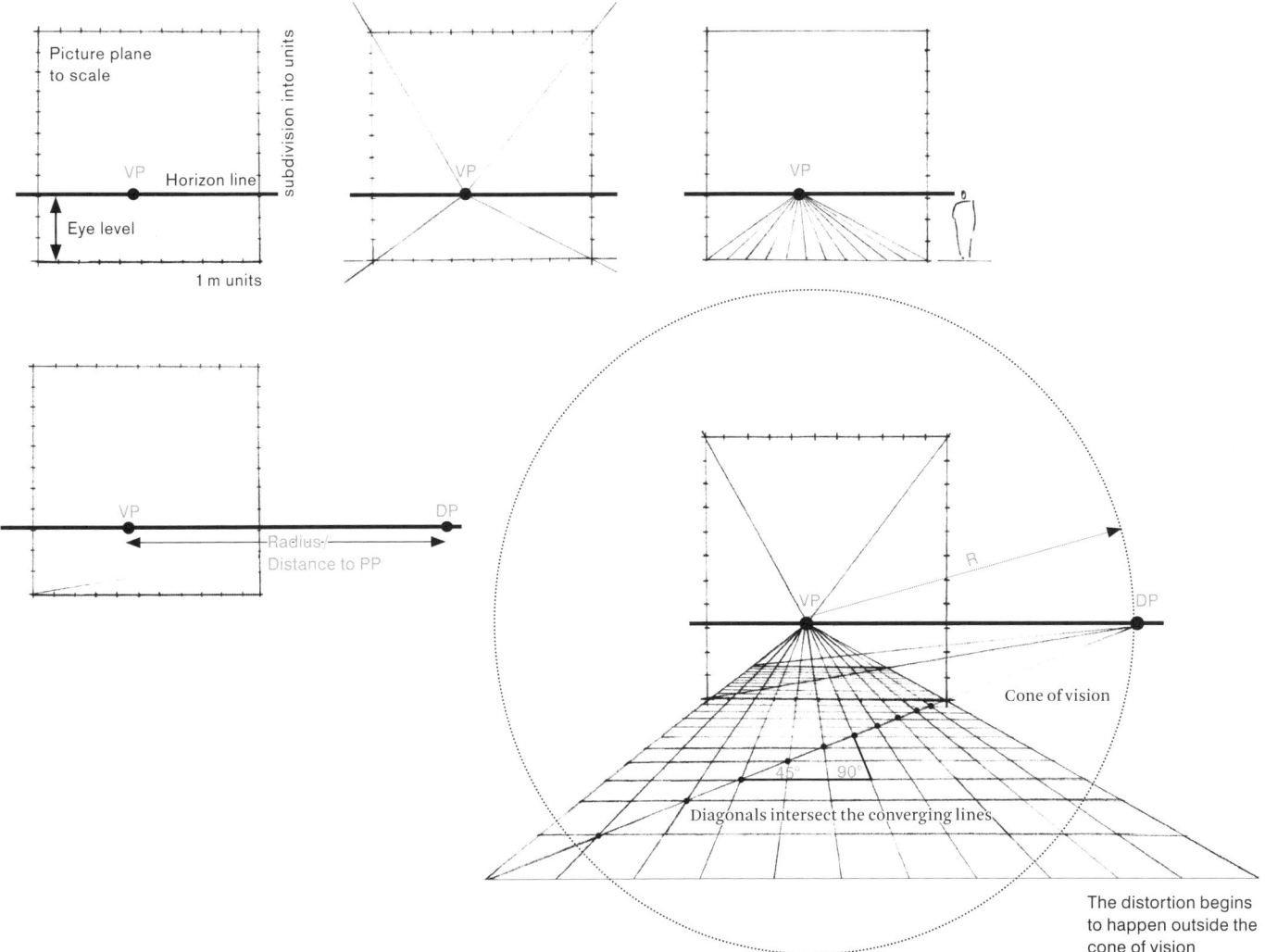

The distortion begins to happen outside the cone of vision

Axonometric projection
Introduction
Characteristics
Vanishing points
Types of perspective
Coordinates and sightlines
Cone of vision
Constructing a perspective grid

Diagonals
Horizon line and pictorial effect
Stairs and ramps
Reflections
Repetitive forms and dimensions
Circles
Simple shadows

6

Diagonal point method for determining depth

The diagonal point method is a quick and easy way to determine scale depth measurement. From geometry, we know that a 45° right angle triangle lies within a square form and will intersect the corners. In the perspective, this means that where the diagonal intersects the converging lines, the square units will occur. The diagonal point for the 45°angle is determined by the distance between the viewer's station point and the picture plane (PP). In a plan view, the line at a 45° angle is simply drawn from the viewer until it intersects the picture plane. This point can be transferred to the horizon line.

The closer the viewer is to the picture plane, the closer the diagonal point will be to the perspective's main vanishing point. If the viewer is further away, some planes may appear foreshortened, even if the overall cone of vision is increased. If necessary to complete the spatial illusion, the depth measurements for the entire space, ground plane, side walls and overhead ceiling can also be established. They can be constructed as a measurable grid extending towards the viewer's horizon. This grid is a helpful basis within which to place objects and to test variations in their locations in a measured space.

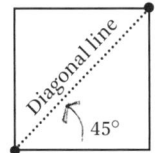

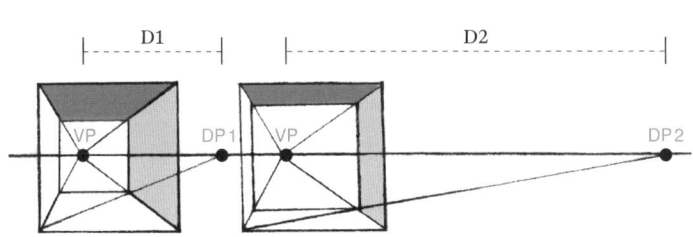

Perspective view with viewer 3m away from PP. The ground plane is clearly visible.

Perspective view with viewer 9m away from PP. The ground plane is now foreshortened as the viewer is further away from the object.

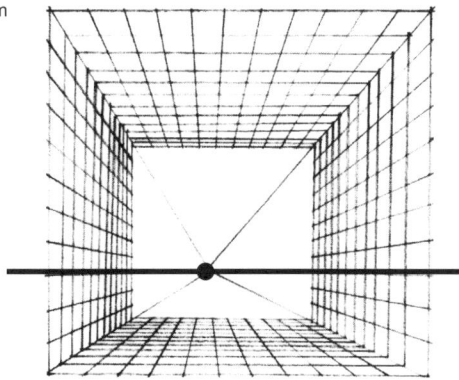

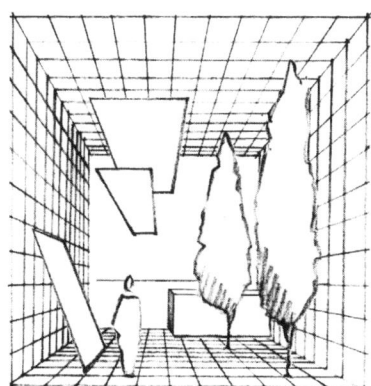

185

Perspective

The horizon line and its effect on the perspective

The horizon line is an imaginary line which we normally do not see in a finished perspective scene. It is however essential for constructing a perspective space and is a key variable for a perspective's set up and for its pictorial effect on the viewer. The examples below demonstrate how the eye level of the viewer, together with the position of the vanishing point can influence how a space is seen and how its parts can be given differing visual emphasis. This is useful in landscape architectural perspectives as it may not always be necessary to include all parts of a space in equal measure.

Horizon at normal eye level / Vanishing point in the middle of the space:
The viewer is standing centrally in front of the object. All sides of the structure are given equal visual emphasis.

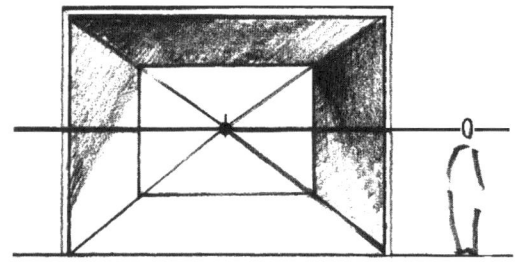

Horizon low / Vanishing point towards the right side of the space:
The viewer is seated towards the right side of the space and his eye level lower than usual. This means that the right wall and ground plane are foreshortened, whereas the ceiling and the left side of the structure appear with greater surface area and are thus given more emphasis.

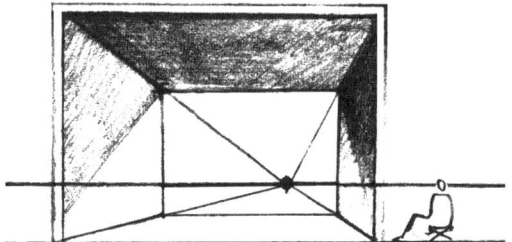

Horizon high / Vanishing point towards the left side of the space:
The viewer is standing above the original ground plane, towards the left side of the space. The left side of the space and the ceiling are foreshortened, whereas the ground plan and the right side of the structure are clearly emphasised.

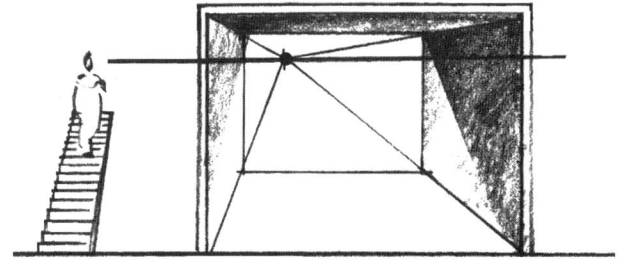

Axonometric projection
Introduction
Characteristics
Vanishing points
Types of perspective
Coordinates and sightlines
Cone of vision
Constructing a perspective grid
Diagonals
Horizon line and pictorial effect
Stairs and ramps
Reflections
Repetitive forms and dimensions
Circles
Simple shadows

6

The height of the horizon line determines the pictorial effect of the perspective view. A perspective drawing can be constructed to include or omit its elements. The viewer can share the same ground plane as the observed scene, be stationed at a lower level or even fly above it, elevating him within the scene to regal heights. Each will determine what is seen the perspective view and will assert an influence on the viewer's perception. The perspective view can be adjusted and even manipulated in order to achieve a deliberate effect on its viewers.

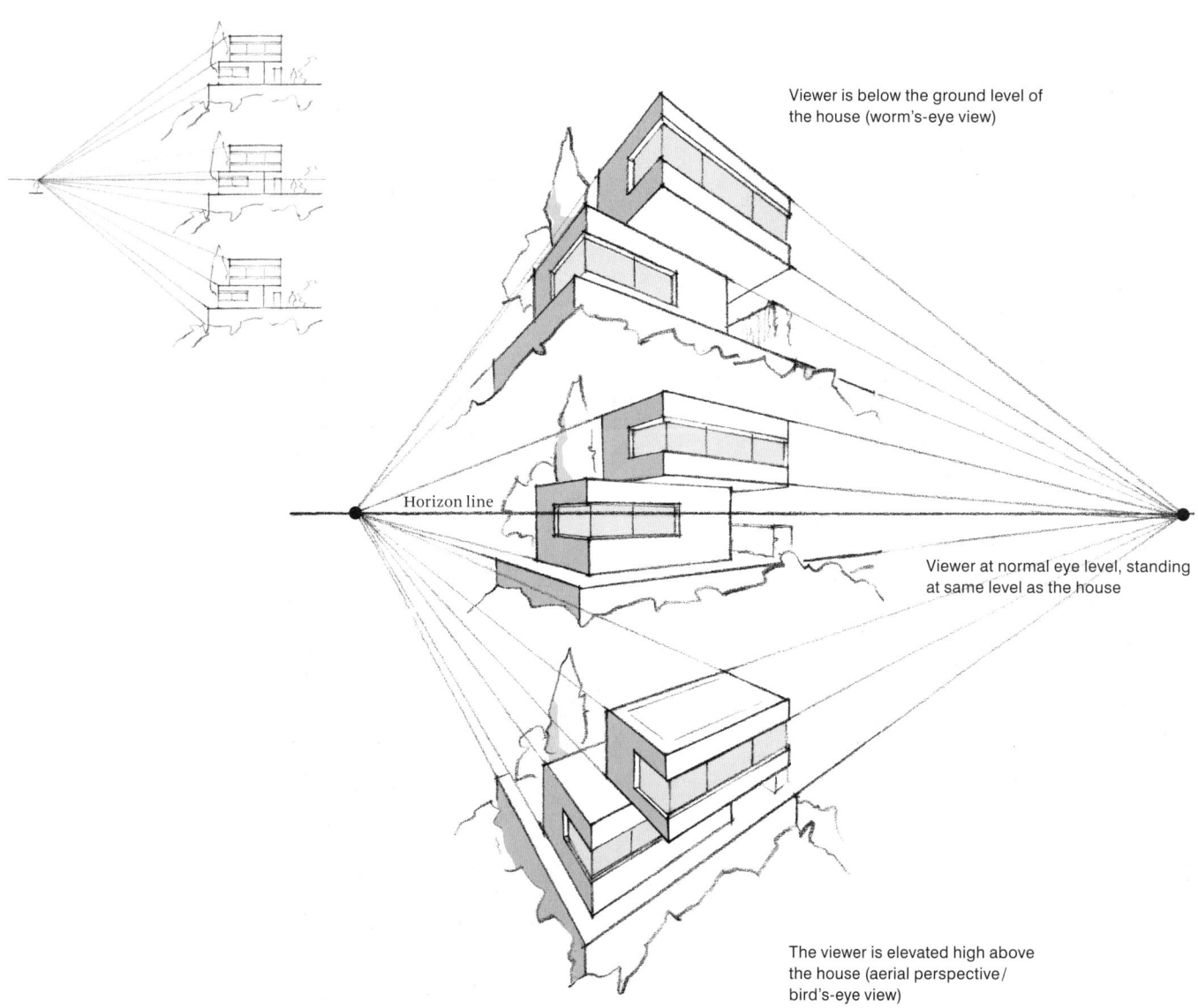

Viewer is below the ground level of the house (worm's-eye view)

Horizon line

Viewer at normal eye level, standing at same level as the house

The viewer is elevated high above the house (aerial perspective / bird's-eye view)

Perspective

Stairs, ramps, and slopes

Stairs and ramps are tilted towards the viewer's picture plane, which means they have their own vanishing points. These will be either above or below the viewer's eye level depending upon direction of the slope's angle. A vertical line drawn through the viewer's vanishing point becomes the vanishing trace line upon which vanishing points of the sloped lines lie. The steeper the slope, the higher up the or down its vanishing point will be along the trace line.

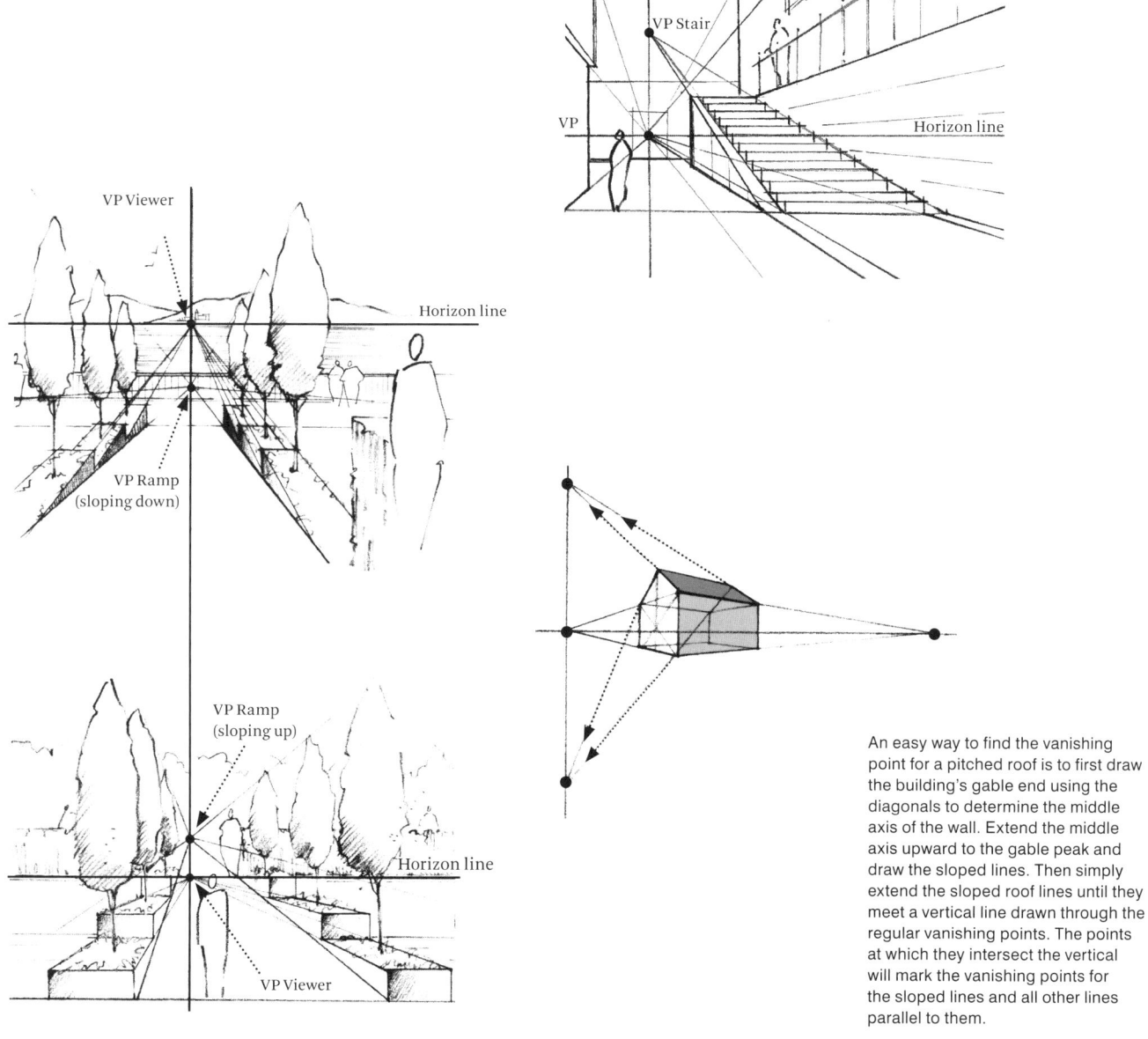

An easy way to find the vanishing point for a pitched roof is to first draw the building's gable end using the diagonals to determine the middle axis of the wall. Extend the middle axis upward to the gable peak and draw the sloped lines. Then simply extend the sloped roof lines until they meet a vertical line drawn through the regular vanishing points. The points at which they intersect the vertical will mark the vanishing points for the sloped lines and all other lines parallel to them.

Axonometric projection
Introduction
Characteristics
Vanishing points
Types of perspective
Coordinates and sightlines
Cone of vision
Constructing a perspective grid

Diagonals
Horizon line and pictorial effect
Stairs and ramps
Reflections
Repetitive forms and dimensions
Circles
Simple shadows

6

Reflections in perspectives

The reflections found in perspective drawings share the same vanishing points as the objects themselves and follow the same perspective rules. Verticals are simply projected from the reflective surface at the same length as the original. If an object is not directly on a reflective plane, the reflection will begin with the same vertical distance to the reflective surface.

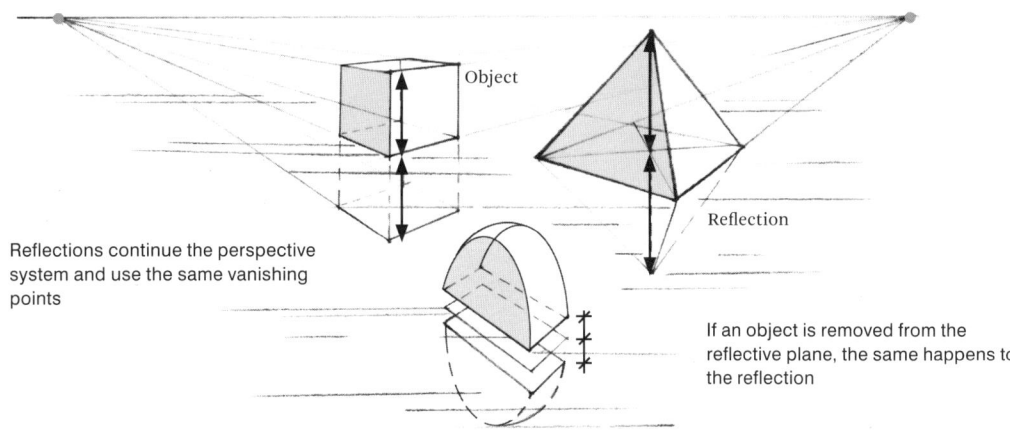

Reflections continue the perspective system and use the same vanishing points

If an object is removed from the reflective plane, the same happens to the reflection

Reflections do not need to be as graphically elaborate or as detailed as the object itself. Water surfaces are drawn using quick horizontal lines, dense at the edges and fading towards the middle, much like water in the plan view.

189

Perspective

Repetitive elements in perspective

Diagonals help to subdivide forms by intersecting in a rectangular or square form to reveal their midpoint. They also help to extend and duplicate repeating planes of the same size in perspective. To demonstrate this principle, we shall look at descriptive geometry and will use a square in elevation.

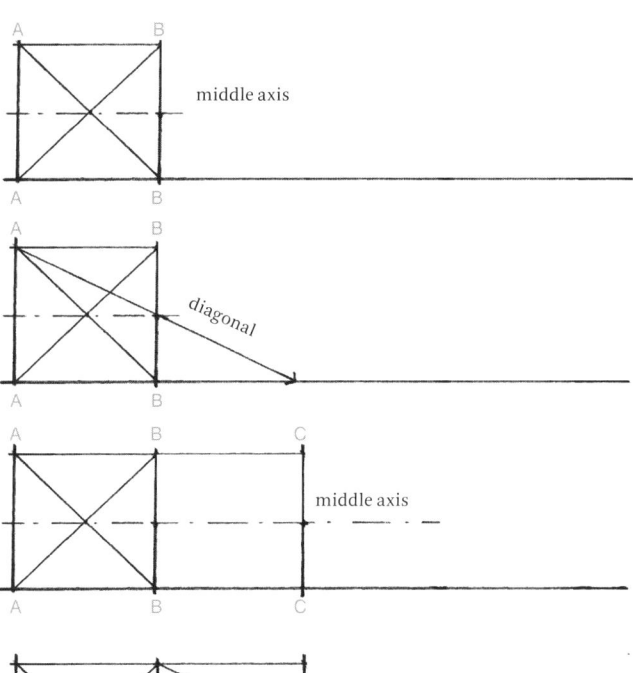

If the distance between the vertical edges AA and BB is to be repeated, two intersecting diagonals are drawn into the plane, revealing the middle point through which a further horizontal axis can be drawn.

Another diagonal is then drawn from the upper corner at AA through the middle of the vertical BB, extending down to the ground plane or baseline.

The point where this diagonal line meets the baseline determines the corner of the next same-sized square. The distance between the vertical lines BB and CC is exactly the same as the distance between AA and BB.

The same method can be used to determine the next same-sized plane and so on. This method is also applicable in the perspective.

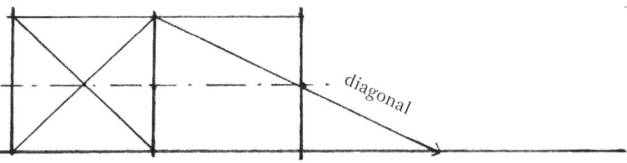

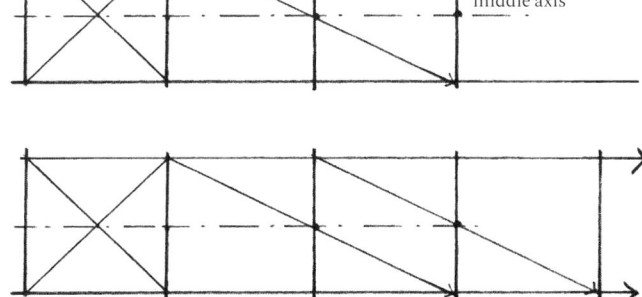

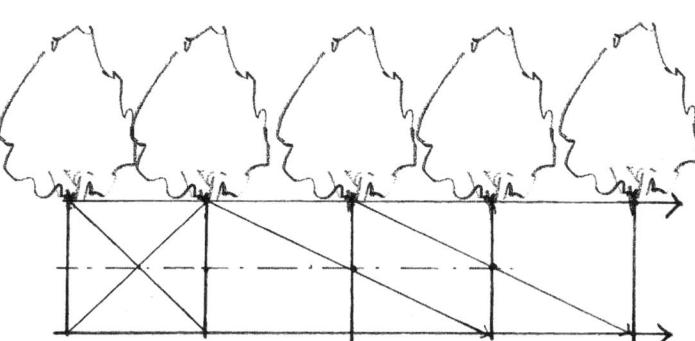

Axonometric projection
Introduction
Characteristics
Vanishing points
Types of perspective
Coordinates and sightlines
Cone of vision
Constructing a perspective grid

Diagonals
Horizon line and pictorial effect
Stairs and ramps
Reflections
Repetitive forms and dimensions
Circles
Simple shadows

6

Foreshortened planes can also be subdivided and repeated using diagonal lines, since two intersecting diagonals in a foreshortened plane still mark the midpoint. This is useful for drawing the gable ends of buildings, where the gable's highest point can quickly be drawn by extending the vertical middle axis of the wall plane.

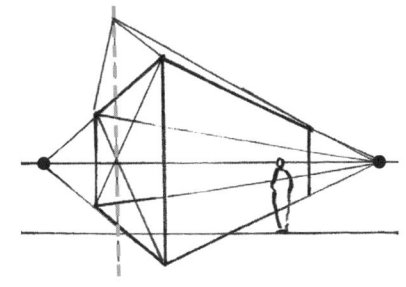

Drawing repetitive elements such as arches and columns in perspective is relatively simple. Once two of the vertical elements are drawn, the rest can be systematically added using only the middle axis and diagonals.

The duplicated vertical columns appear increasingly foreshortened and smaller in size as they get closer to the vanishing point.

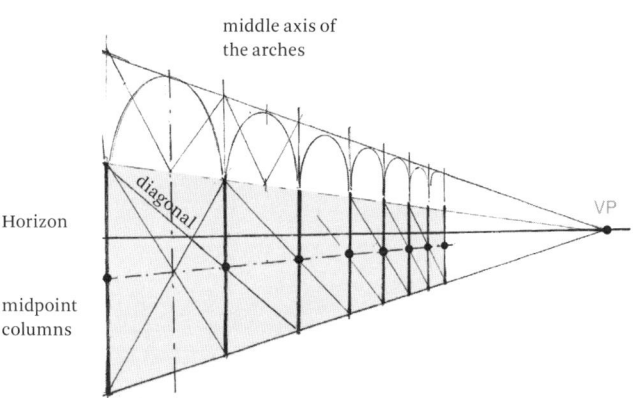

This simple method saves a lot of time and estimation, especially in freehand perspective drawing. It can help to draw repeating vertical elements, such as allees of trees or rows of columns. This method can also be applied to horizontal planes, for example when drawing paving patterns on the ground surface.

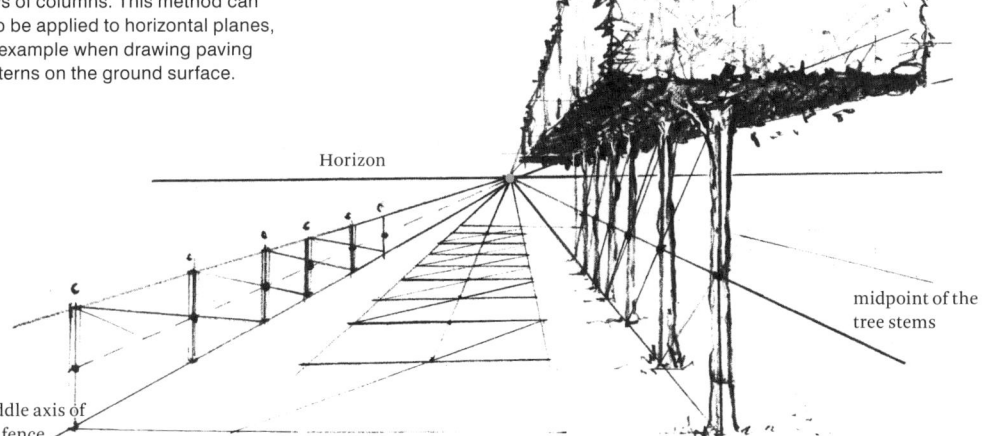

191

Perspective

Drawing circles in perspective

Circles only remain circles if they are parallel to the picture plane. On foreshortened planes, they become ellipses. They are easily drawn when circumscribed with a square and further subdivided using diagonals. Using the points where the circle intersects the different divisions can aid the construction of the ellipse in the perspective.

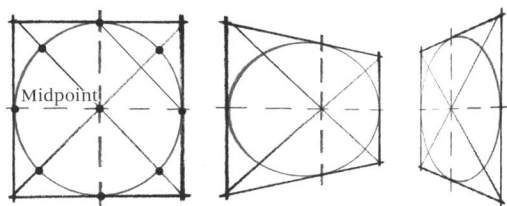

Subdivisions help to locate the ellipse within the foreshortened plane

This sketch below shows the circular hedges with conical trimmed plants. Once the ellipses were drawn on the ground, the conical plants were easily constructed. The midpoint of the circle was vertically extended to the desired height to form the conical shape.

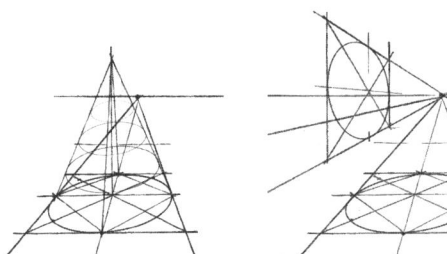 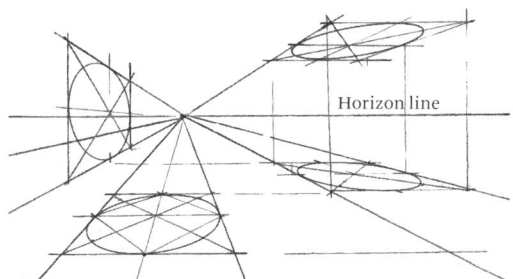

Axonometric projection
Introduction
Characteristics
Vanishing points
Types of perspective
Coordinates and sightlines
Cone of vision
Constructing a perspective grid
Diagonals
Horizon line and pictorial effect
Stairs and ramps
Reflections
Repetitive forms and dimensions
Circles
Simple shadows

6

Constructing simple shadows

Shadows are a complex subject and are only briefly explained here. These are three simple ways to construct shadows in a perspective. Adding a light source will add volumes and three-dimensionality to a scene. Shadow shapes will depend on the shape of the object, but also on the type and position of the light source itself. Shadow forms share the vanishing points of the objects in the image, but they also have their own vanishing point in the bearing direction of the radial light rays.

Light source behind the picture plane, in front of viewer:

The light source is well above the horizon. It sends out radial light rays which touch the corners of the object and end at the ground plane.

From these points, the shadow form converges towards vanishing points, on the ground plane. One usually lies directly below the light source, in the so-called bearing direction.

The points are connected to form the shadow planes and given a tonal value to appear darker than the object.

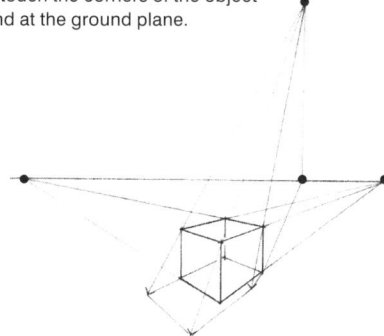
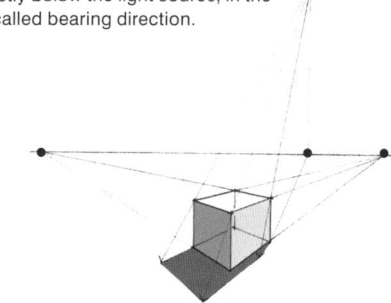
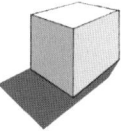

Light source parallel to the picture plane, to the side of the viewer:

The light source here is to the right of the viewer, sending out parallel light rays. They touch the edges of the object and extend to the ground plane.

The parallel edges stay horizontal and do not converge at a vanishing point.

This shadow construction is quick and easy, although the resulting shadows may appear less dynamic than the previous example.

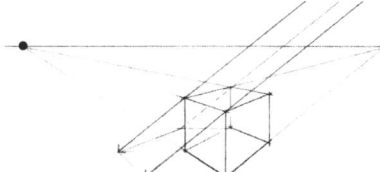
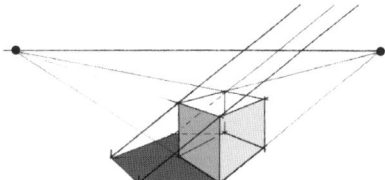
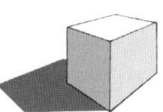

Light source in front of the picture plane, behind the viewer:

The light source is behind the viewer and has its vanishing point below the horizon line. The light rays have their own bearing direction which extends downwards from the object edge.

Since shadows can be very complex constructions, it is sufficient to keep them simple in a sketch drawing. Adding shading and diffuse shadows

which fade into the ground plane are often enough to suggest a light source and add volume. More complicated shadows are usually left to CAD.

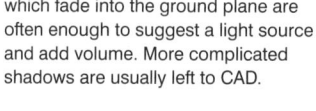
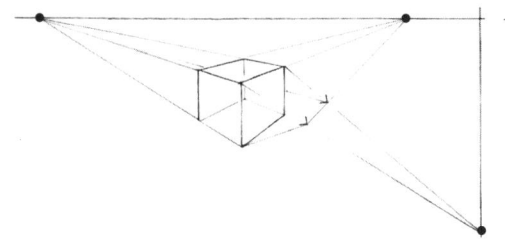
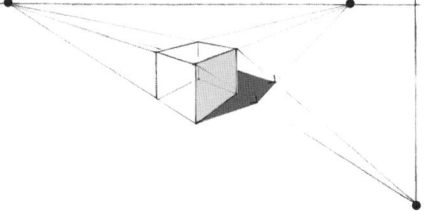
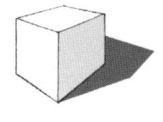
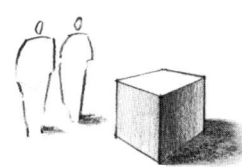

Perspective

Perspective plan method: One-point perspective
A common method, also called the office method, with which to construct a perspective is to project it directly from a plan. Here, the viewer is placed in the plan at a fixed station point, with the subject of interest within his cone of vision. The picture plane is then located in the plan, usually passing through or along an object at a significant vertical point. Directly above the plan view, a baseline (the ground plane) and a horizon line at the desired eye level can be drawn. The central sightline of the viewer is projected onto the picture plane and vertically upwards to form the central vanishing point on the horizon line. Unlike other projections, perspective views are not to scale. The picture plane in the plan view does function as measuring device, as what ever it touches can be drawn to a scaled height in the perspective. The steps for constructing a one-point perspective from plan are briefly demonstrated here using a simple cube, with one side parallel to the picture plane.

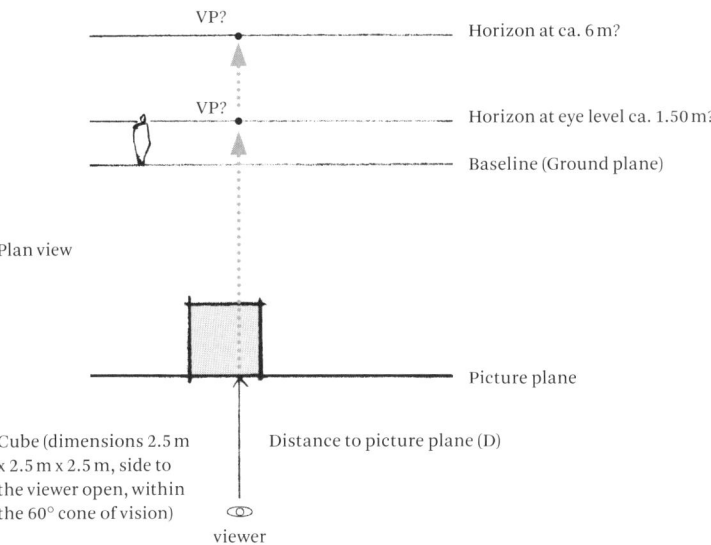

Perspective view

Horizon at ca. 6 m?

Horizon at eye level ca. 1.50 m?

Baseline (Ground plane)

Plan view

Picture plane

Cube (dimensions 2.5 m x 2.5 m x 2.5 m, side to the viewer open, within the 60° cone of vision)

Distance to picture plane (D)

viewer

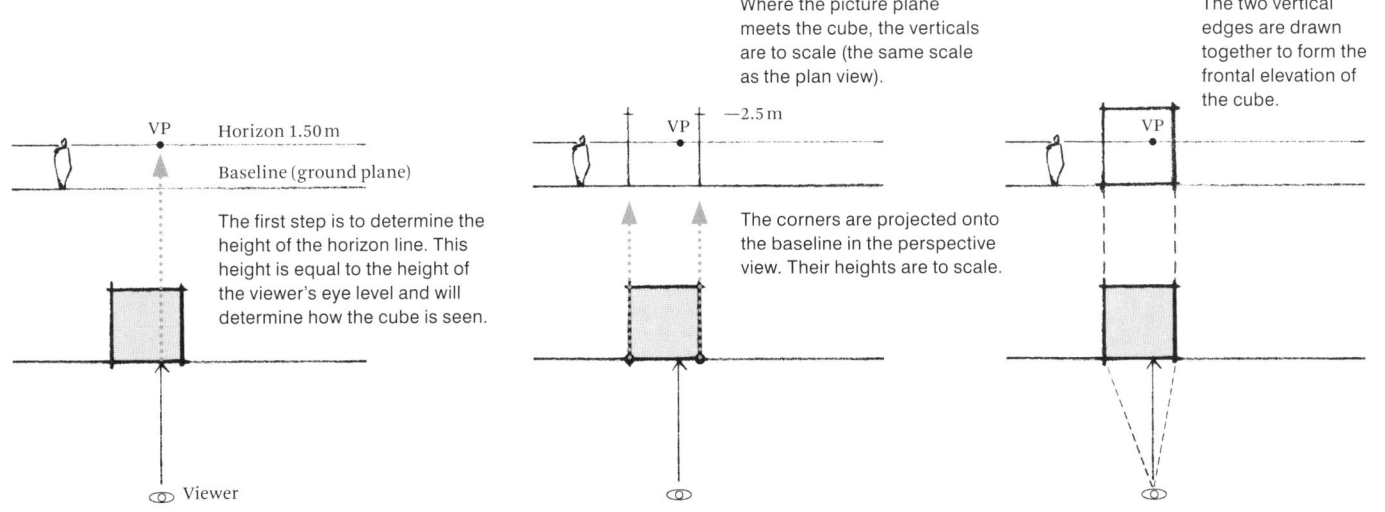

VP Horizon 1.50 m
Baseline (ground plane)

The first step is to determine the height of the horizon line. This height is equal to the height of the viewer's eye level and will determine how the cube is seen.

Viewer

Where the picture plane meets the cube, the verticals are to scale (the same scale as the plan view).

−2.5 m

The corners are projected onto the baseline in the perspective view. Their heights are to scale.

The two vertical edges are drawn together to form the frontal elevation of the cube.

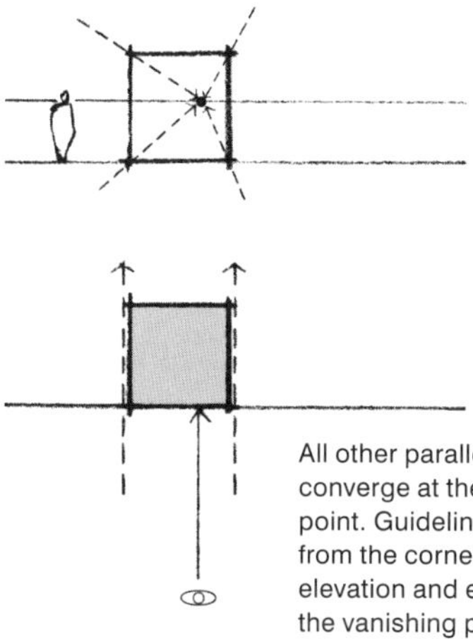

Construction methods
From the plan view
Using a perspective grid
From photos
Drawing freehand perspectives
Estimating proportions
Freehand one-point perspective
Freehand two-point perspectives
Atmospheric perspective
Graphic emphasis

6

All other parallel lines converge at the vanishing point. Guidelines are drawn from the corners of the front elevation and extended to the vanishing point on the horizon line. They appear to go on infinitely.

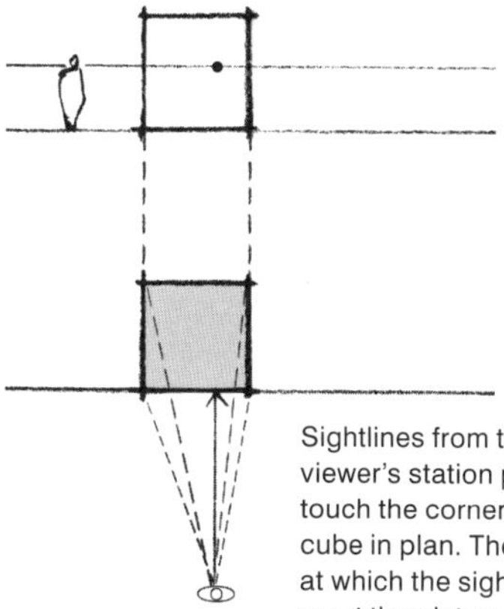

Sightlines from the viewer's station point touch the corners of the cube in plan. The point at which the sightlines meet the picture plane are marked.

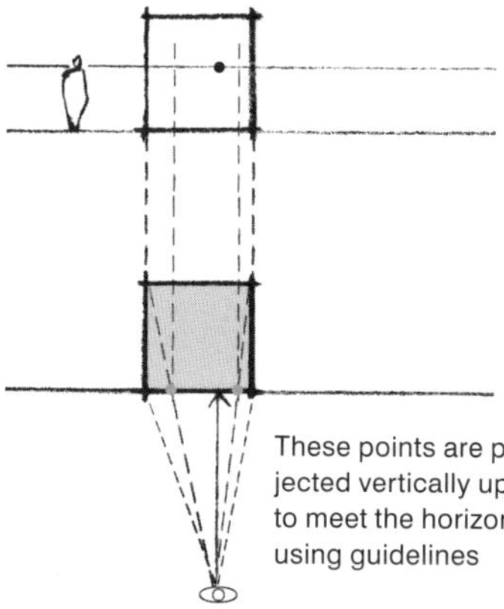

These points are projected vertically upwards to meet the horizon line using guidelines

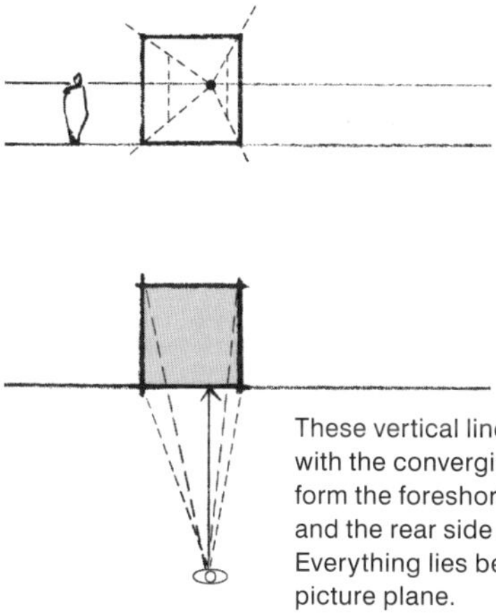

These vertical lines intersect with the converging lines to form the foreshortened sides and the rear side of the cube. Everything lies behind the picture plane.

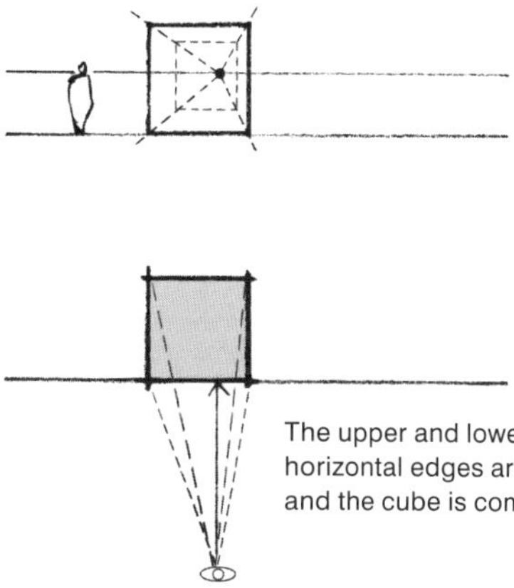

The upper and lower horizontal edges are drawn and the cube is complete

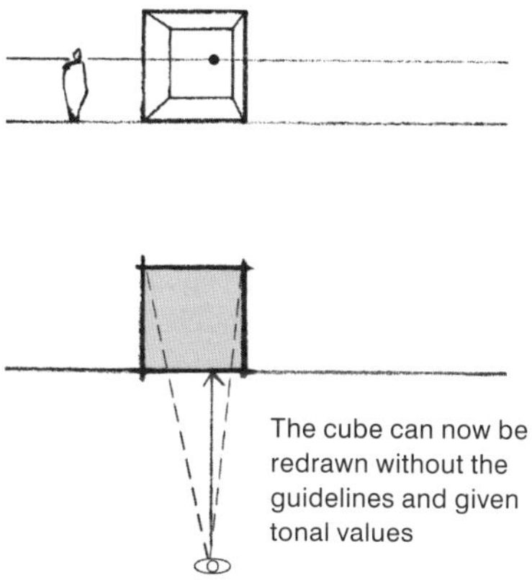

The cube can now be redrawn without the guidelines and given tonal values

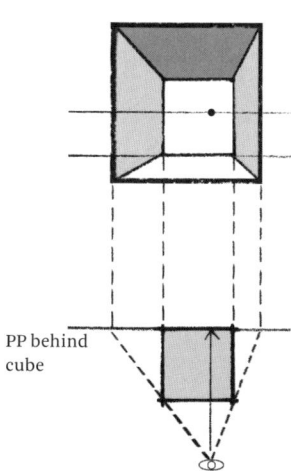
PP behind cube

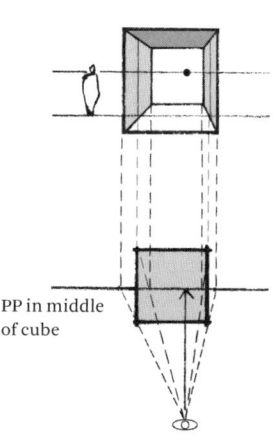
PP in middle of cube

The picture plane determines the size of the object in the perspective image. Objects placed in front of it will appear larger to the viewer, whilst objects behind will appear smaller.

Even though the viewer's station point is unchanged, these examples show how a perspective view can be altered depending on the location of the picture plane in plan.

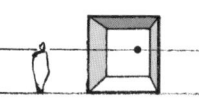

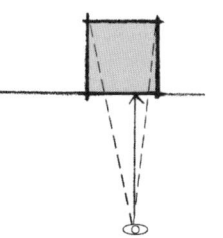
PP in front of cube

195

Perspective

Perspective plan method: Projecting a two-point perspective
Here another very simplified demonstration showing how a simple volume – in this case the same cube – can be drawn in two-point perspective using a plan. This time its position is rotated towards the viewer, with no sides parallel to the picture plane.

Perspective view

Plan view

Cube (2.5 m x 2.5 m x 2.5 m), rotated, side to the viewer open, within the 60° cone of vision

Horizon line 1.50 m
ground plane (viewer)

picture plane

viewer

The vanishing points for the cube lie to the left and right of the viewer. Their location is determined by the angle of rotation. Two lines parallel to the rotated cube are drawn from the viewer's station point to the picture plan. The points of intersection are then projected vertically to the horizon in the perspective view above.

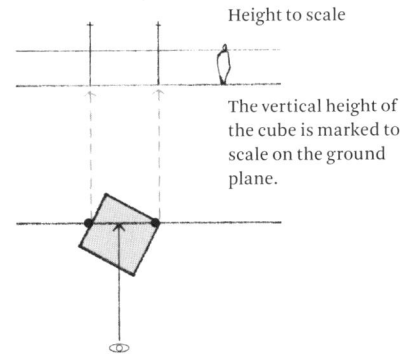

Height to scale

The vertical height of the cube is marked to scale on the ground plane.

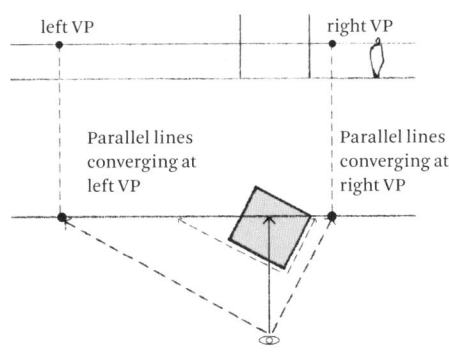

left VP   right VP

Parallel lines converging at left VP

Parallel lines converging at right VP

The points of intersection between the cube and the picture plane can be used as vertical measuring lines in the perspective view (located here above the plan). They are to scale. It is always a good idea to place the picture plane through a significant vertical edge of an object, such as a corner. Sightlines are drawn from the viewer's station point to the cube's corners. The point where a sightline intersects the picture plane will be projected vertically upwards into the perspective view where it represents a vertical line of the cube's edge. Where the corner lies in front of the picture plane in plan, the sightline is simply extended to reach the picture plane, and this point is then projected upwards.

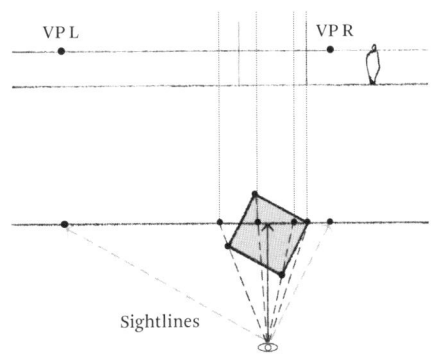

VP L   VP R

Sightlines

Construction methods
From the plan view
Using a perspective grid
From photos
Drawing freehand perspectives
Estimating proportions
Freehand one-point perspective
Freehand two-point perspectives
Atmospheric perspective
Graphic emphasis

6

With the help of the first two measured heights drawn on the ground plane line, the two sides of the cube can now easily be drawn. Using these heights as a reference, their measurements are extended towards each of the vanishing points, both right and left.

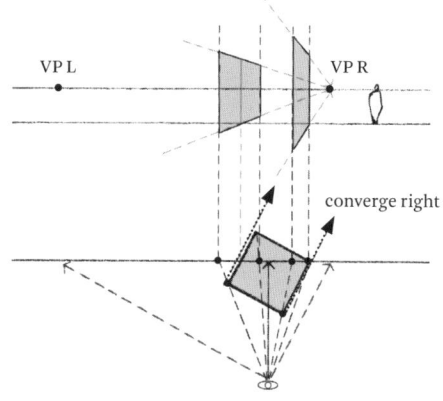
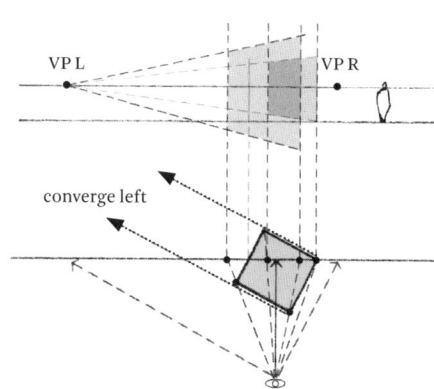

If the two sets of planes are put together, it emerges the cube in perspective. In a final drawing, the cube is redrawn without the construction lines and rendered using light and shade.

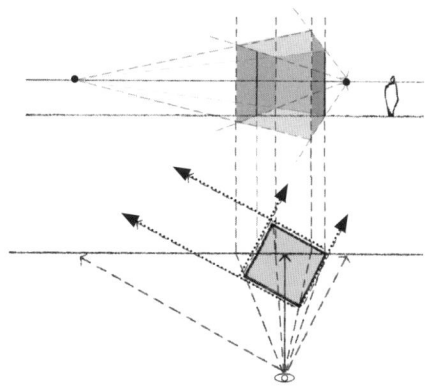
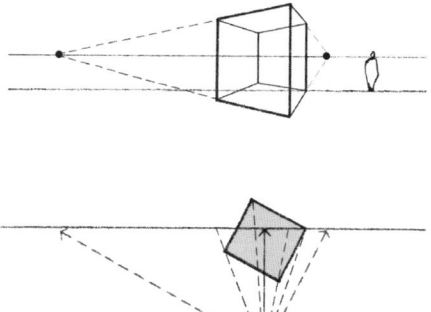

The location of the picture plane in the plan not only offers a measured height to start with, it also will determine the size of what is shown. Similar to the example on the preceding pages, these two-point perspectives show how the size of an object in the perspective is affected by the picture plane in plan.

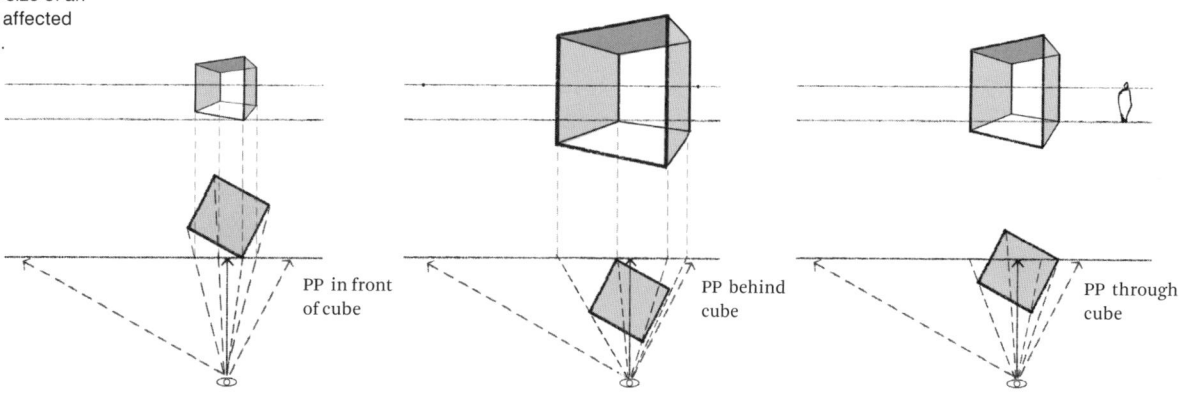

Perspective

Drawing perspectives with a prepared perspective grid
Using a prepared perspective grid, also called a perspective chart, is a quick way to test ideas without using a plan view. We can work directly in the coordinate system.

Perspective variables like eye level, station point and scale are all pre-determined and no longer flexible. All objects and plane must adhere to the perspective principles.

In the left column are two one-point perspective grids. In both examples the viewer is stationed directly in the middle of the picture, but at different eye levels. In the drawing above the viewer is on the ground plan with an eye level of 1.6 m, while in the drawing below he is hovering above the ground at a height of 8 m. The vertical measuring line (subdivided into units) is shown in the distance.

In this column we have two-point perspective grids, also at different eye levels. Along with the vertical measuring line, the cone of vision is indicated here.

Eye level 1.60 m

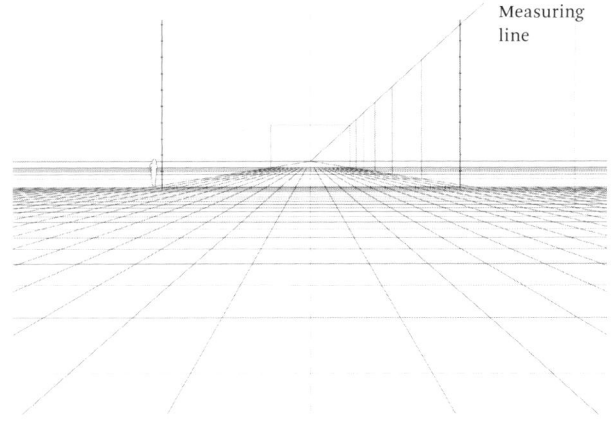

Eye level 1.60 m

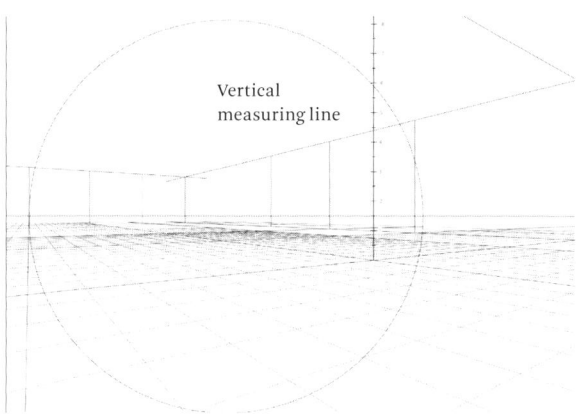

Eye level 8.00 m

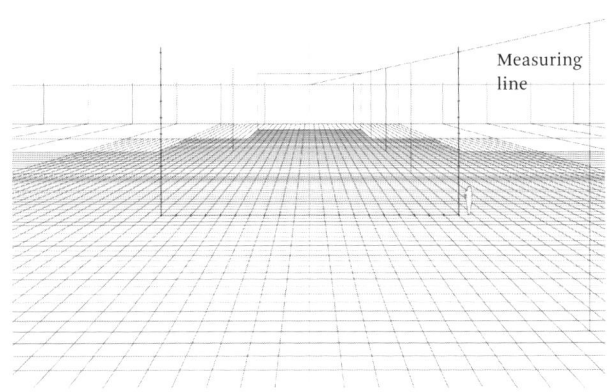

Eye level 8.00 m

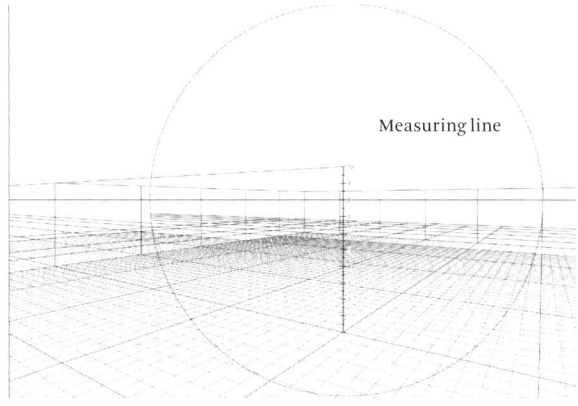

Construction methods
From the plan view
Using a perspective grid
From photos
Drawing freehand perspectives
Estimating proportions
Freehand one-point perspective
Freehand two-point perspectives
Atmospheric perspective
Graphic emphasis

6

With good knowledge of perspective principles, it is easy to use perspective grids to construct and test design ideas and graphic variations.

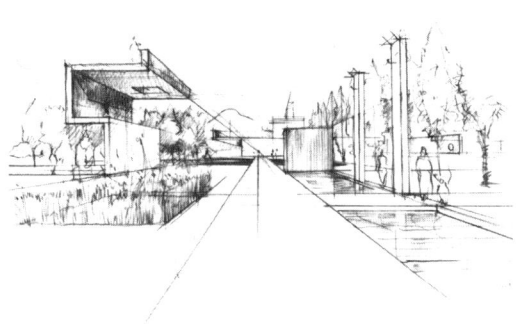

Perspective

Using a perspective grid

Using a prepared grid is a handy way to sketch and communicate a spatial experience without referring to a finished plan. In this example, the viewer is in the middle of the picture plane, looking forward at an eye level of 1.60 m. The vertical measuring line is subdivided into increments of 1 m, both vertically and horizontally. The ground plane is fully rendered with 1 m x 1 m square units. The viewer is standing in the middle of the scene and it is best if his station point is at the very front of the grid so as to maximise the larger squares directly in front of him. This station point will correspond to his location in plan, should a plan be used. Once this is established, objects and surfaces can be transferred into the perspective space. Vertical heights can be measured along the vertical measuring edge. Heights and widths are to scale only along these lines, However, if the measuring line lies far back in the space, there is a trick for vertical estimations which only works in a one-point perspective grid. The horizontal length of a square on the ground plane can be used to find the equivalent vertical height, simply by rotating the same length by 90° into a vertical position. More units of a similar length can be added to achieve a desired height.
The 1 m x 1m squares can be further subdivided, for example by using diagonals to find their midpoints.

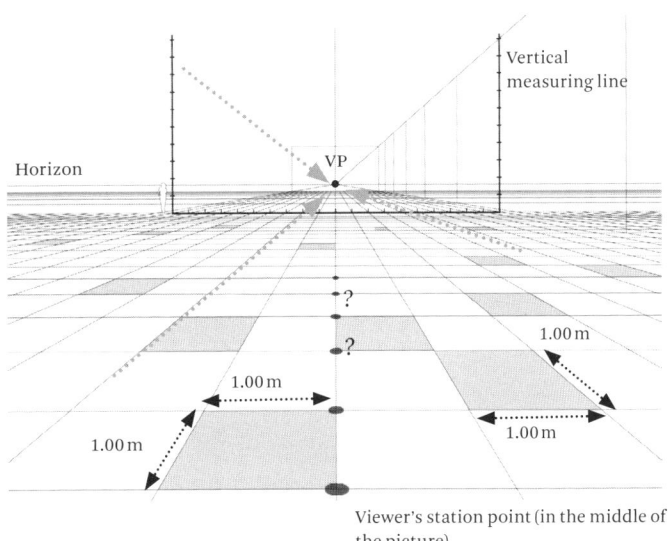

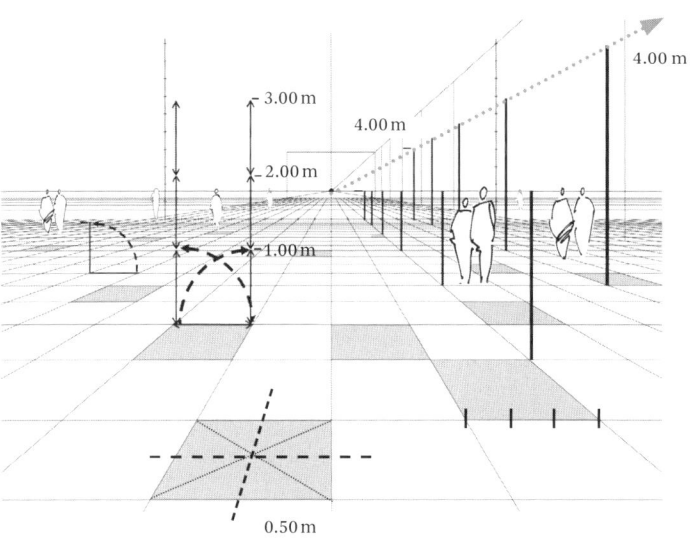

Vertical heights can be determined by simply rotating a known horizontal length 90° vertically from the corner of the line. This will have the same unit of length as its horizontal counterpart at that point on the ground plane.

Measurements taken on the vertical measuring line can be projected forward or backwards from the line

The square can quickly be divided using intersecting diagonals

Further subdivisions along the horizontal lengths of the square units, can be estimated

Construction methods
From the plan view
Using a perspective grid
From photos
Drawing freehand perspectives
Estimating proportions
Freehand one-point perspective
Freehand two-point perspectives
Atmospheric perspective
Graphic emphasis

Planes and surfaces parallel to the picture plan stay parallel. They reduce in size as they get closer to the vanishing point, appearing more distant to the viewer. Planes and surfaces which do not lie parallel to the picture plane converge towards the vanishing point, foreshortening along the way. Since circles can be circumscribed by a square, it is relatively easy to construct them in a perspective grid full of squares. It is still recommendable to draw diagonal subdivisions in order to fit the ellipse into the foreshortened square plane. People always enliven a scene, especially so in a perspective. A quick way to draw people in a perspective scene is to draw a few people about the same size as the viewer. This means that their heads will lie around the horizon line, although their overall size will vary according to their location on the ground plane. Larger and smaller figures can then be added as necessary. Generally speaking, it is easier to construct orthogonal designs in a perspective grid. However, irregular shapes can also be transferred to a perspective grid by using the point where the shape intersects with the grid. Stairs and ramps can also be drawn in the perspective grid, however this method works best for designs with minimal level changes.

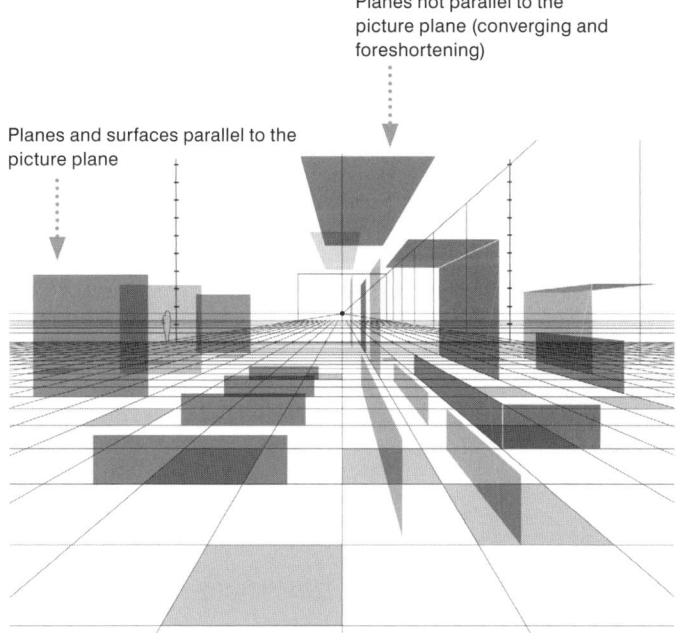

Foreshortened planes add convincing visual depth to a scene.

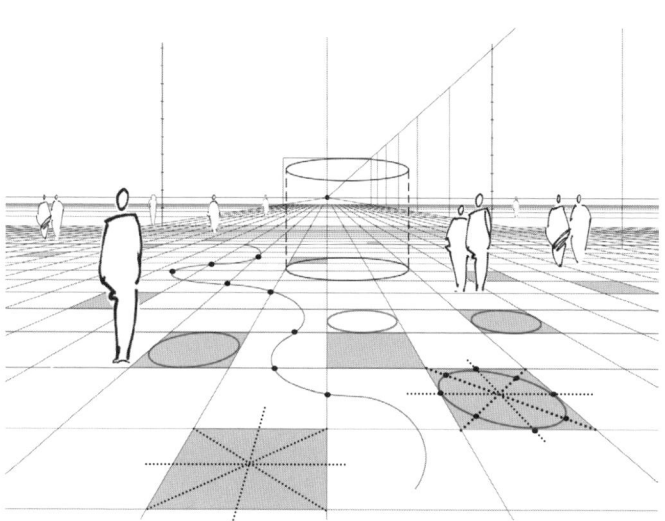

Diagonals are helpful when constructing an ellipse into a square.

Perspective

Using a prepared perspective grid
The process of transferring a plan view into a perspective grid is demonstrated here, using a small garden plot.

The first requirement is that the garden plan must be to scale and overlaid with a grid.

In this example, the viewer is standing in the middle of both the picture plane and the middle axis of the garden itself. This means that both sides of the space get equal graphic emphasis in the perspective. The viewer's location, a short distance in front of the garden, ensures that its important elements lie within the 60° cone of vision.

Had the viewer been placed towards the left or right of the garden, or further into the garden space itself, the perspective scene would have highlighted different elements (see page 186). Regardless of position, one side of the garden must be parallel to the picture plane.

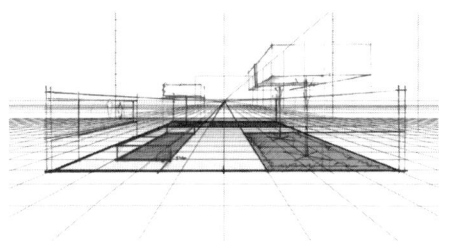

Vertical dimensions are measured along the measuring line and brought forward into the garden area. Basic elements can be constructed and drawn at their locations.

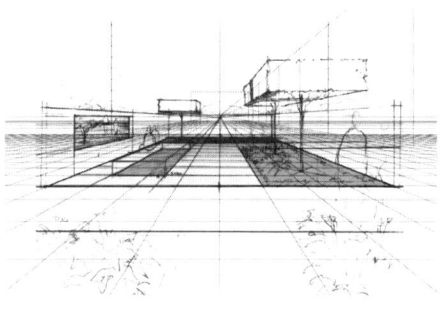

When all parts are constructed, everything is redrawn and rendered on an overlay in ink, leaving out construction lines.

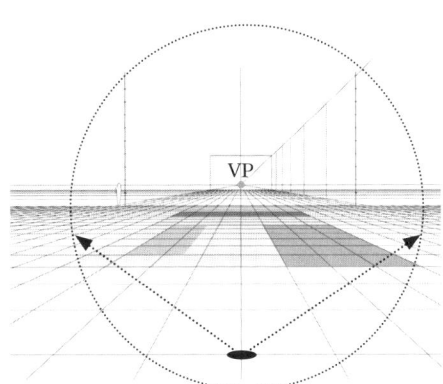

With the help of the station point in plan, the garden's outline, internal areas and divisions are transferred into the grid metre by metre.

People and vegetation enliven the scene. After rendering, the drawing was also scanned and a sky was added.

Construction methods
From the plan view
Using a perspective grid
From photos
Drawing freehand perspectives
Estimating proportions
Freehand one-point perspective
Freehand two-point perspectives
Atmospheric perspective
Graphic emphasis

6

In an aerial perspective, both the garden and its immediate environment become visible. It is a good way to let the viewer understand the area which directly surrounds the garden, along with its overall context, whether urban or rural.

Viewing a site from above may well distance the viewer from the property, however the resulting effect is often more impressive than the experiential perspective at eye level.

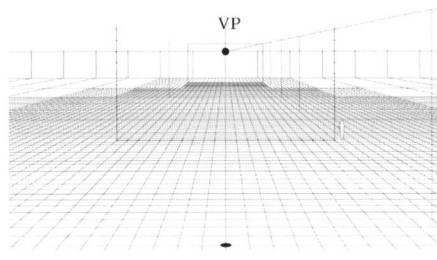

In this example, the viewer is still central in the scene, however he is hovering 8 m above the garden. The immediate surroundings are as visible as the garden.

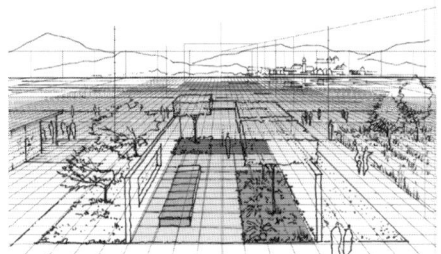

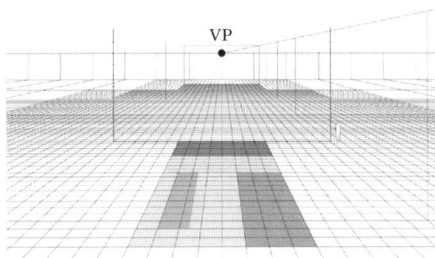

The garden's outline and ground plane (here a little longer than the previous example) are transferred metre by metre into the grid.

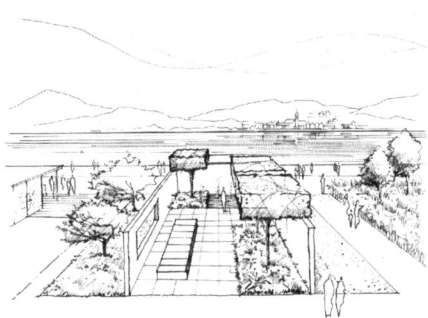

The constructed drawing is overlaid and redrawn in ink, using rendering techniques to graphically enliven the scene.

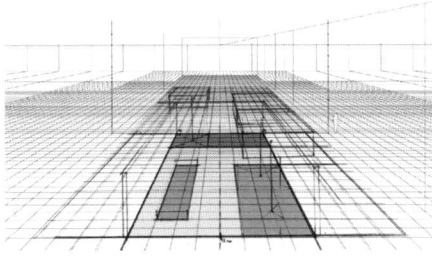

Vertical dimensions are measured along the measuring line and brought forward into the garden space. Important forms and garden elements are constructed along with basic context.

The garden and its context are rendered and detailed for the final presentation.

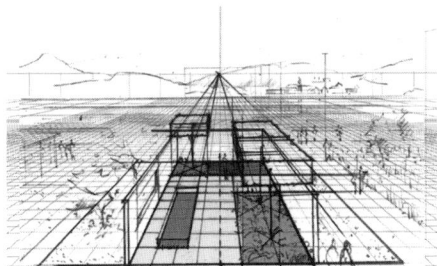

203

Perspective

Drawing perspectives using photographs

 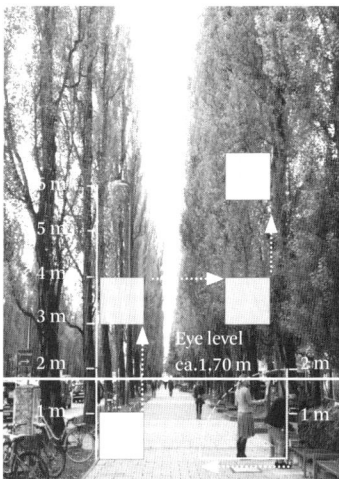 

A photograph can be an excellent base to start sketching perspectives. Design variations can be quickly tried out directly in their context. The first step is to determine the horizon line, which will usually be at the eye level of the photographer. By tracing the converging lines, it is easy to establish the vanishing point.

Vertical heights will have to be estimated using existing vertical elements and objects in the image. These can be people or street furnishings. Known lengths in the horizontal ground plane can also be rotated to find their vertical equivalents.

Converging lines can be quickly drawn towards the vanishing point. Sizes and proportions towards the vanishing point will have to be estimated.

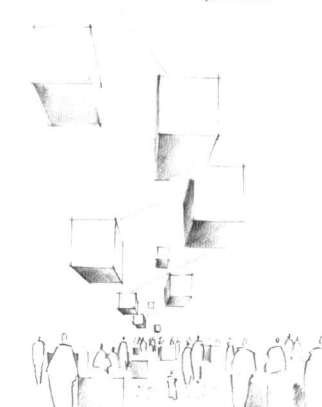 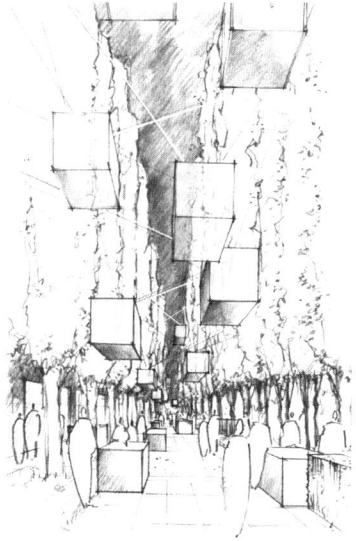

By overlaying a photo with sketch paper, visual compositions and design variations can quickly be tested out. These perspectives can be redrawn entirely by hand or converted into collages of both photo and manual rendering.

Construction methods
From the plan view
Using a perspective grid
From photos
Drawing freehand perspectives
Estimating proportions
Freehand one-point perspective
Freehand two-point perspectives
Atmospheric perspective
Graphic emphasis

A similar method can be applied to a two-point perspective photo. First, the horizon and the vanishing points are established. It is common for one or both of the vanishing points to lie outside the photograph.

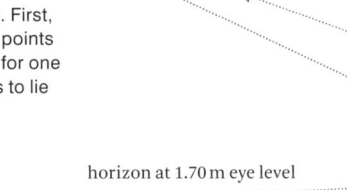

With the help of vertical elements in the photo (in this case the corner of the barn structure), vertical heights and dimensions can be estimated. These vertical markers can then be transferred into the perspective space using both vanishing points.

The photo can now be overlaid with sketch paper, redrawn and enhanced with design elements, vegetation and people. Adding light and shade effects, whether manually or digitally, adds depth to the composition.

Perspective

Freehand perspective drawings on site

In order to sketch a perspective freehand on site, it is important to establish one's own horizon line within the seen space. Only then can the perspective be constructed. This does not mean that one begins with a horizontal line on the drawing surface. A good way to start a perspective drawing is by sketching a vertical reference: this could be either a corner or even a façade within which one can estimate one's eye level. Once the basic proportions of the façade are drawn, it helps to visualise oneself standing directly in front of it or at the corner of the building. It might be a good idea to send a similar-sized friend over to the building, in order to "see" one's eye-level, as demonstrated by this person. Once the height is established within the overall proportion of the structure, the vanishing point can be determined. Once the vanishing point is drawn, constructing the perspective is straightforward. Parallels perpendicular to the picture plane (or drawing surface) converge towards the vanishing point. These can be path edges, rows of trees or walls. As the drawing builds up, the construction lines and even the horizon line can be erased. Do not be afraid to leave them in or overlay lines if it helps you remember where important parts of the perspective are.

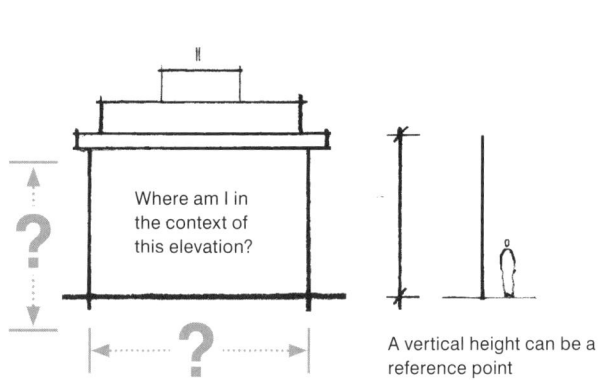

Where am I in the context of this elevation?

A vertical height can be a reference point

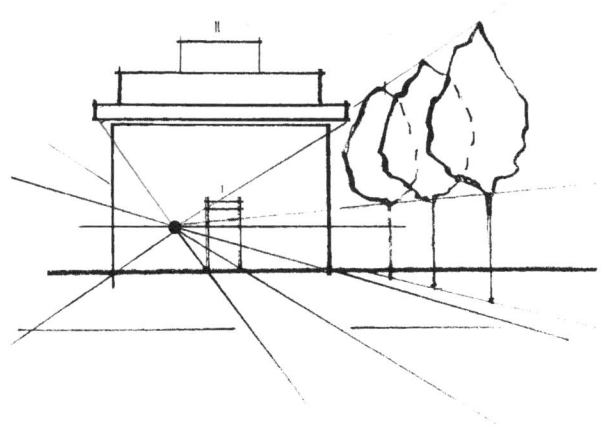

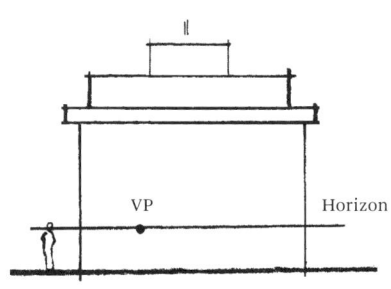

VP   Horizon

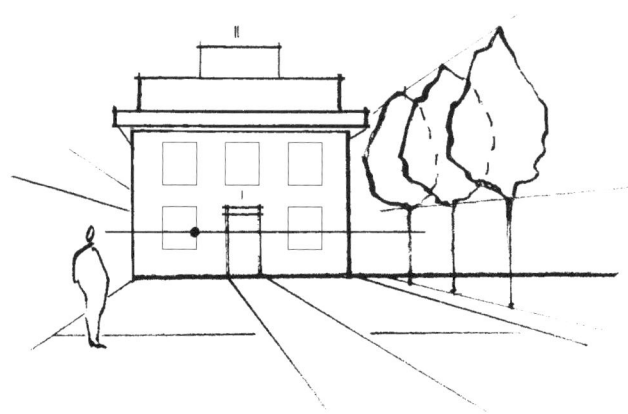

Construction methods
From the plan view
Using a perspective grid
From photos
Drawing freehand perspectives
Estimating proportions
Freehand one-point perspective
Freehand two-point perspectives
Atmospheric perspective
Graphic emphasis

The same principle can be applied to the two-point perspective. The first step is to find the horizon line in proportion to a vertical height. One place to start may be a building corner. This vertical marker can be put down on paper. The horizon line, the viewer's personal eye level, is determined along this line and then drawn. The two vanishing points will lie on this line; their location will have to be judged carefully in accordance with what one sees. Converging lines can then be drawn towards their respective vanishing points and the perspective carefully constructed. When drawing a freehand perspective it is important to make sure that the eye level or horizon line is not positioned too high in the scene. Always ensure that the eye level in the drawing corresponds to your own height before the perspective is fully constructed and rendered. If the perspective shows views over the rooftops, even though you are standing firmly on the ground, then the horizon line was positioned incorrectly.

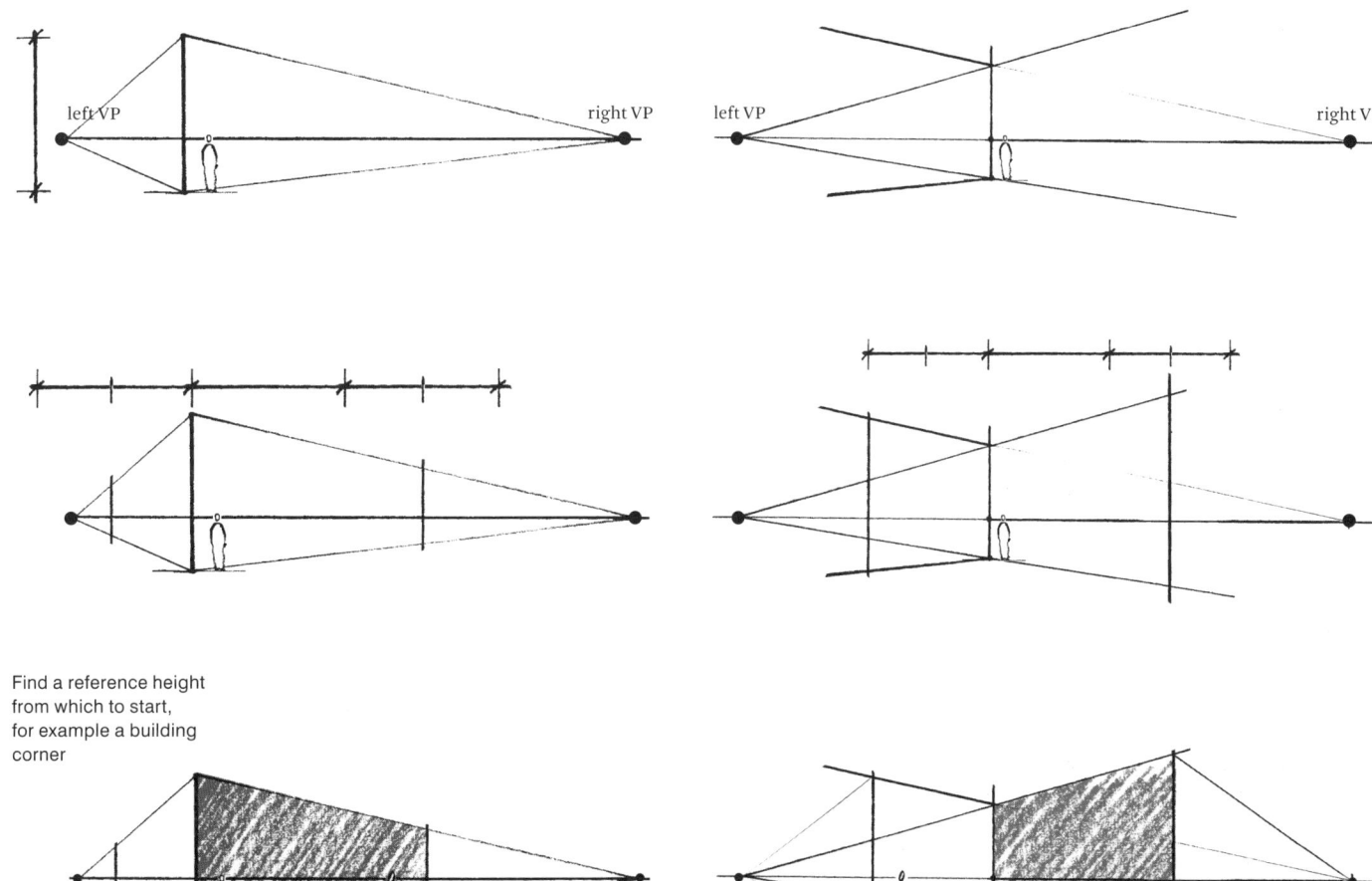

Find a reference height from which to start, for example a building corner

Perspective

Estimating proportions when freehand drawing
Horizontal and vertical proportions can be quickly gaged using just the pencil or pen. The technique is explained by the drawings on these pages.

Hold the pencil upright and stretch out your arm in front of you. Look at both the object and the pencil using only one eye.

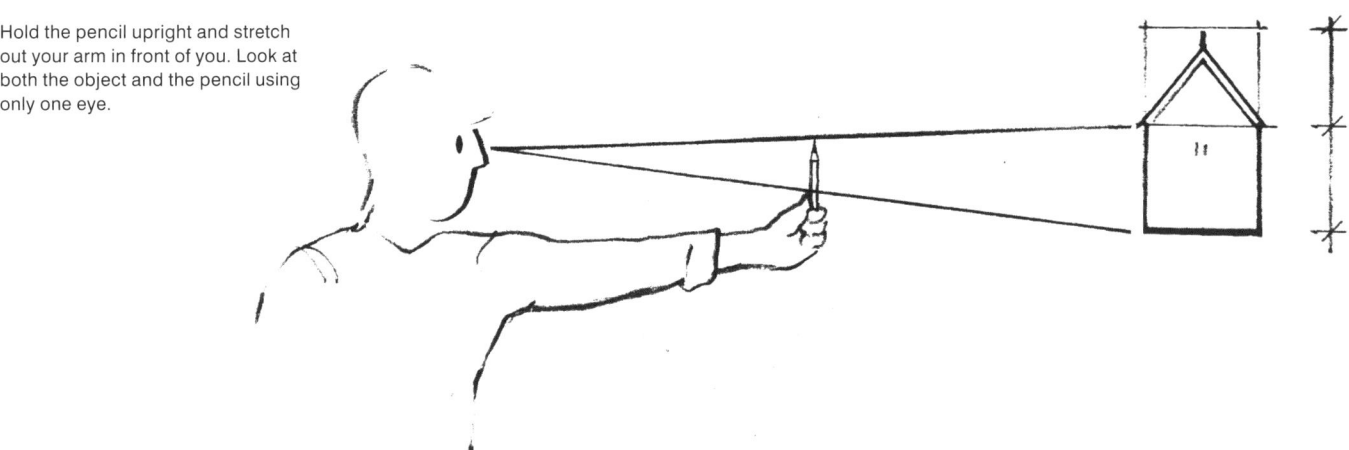

Once again, we have to begin with some point of reference. This is often a measuring line, either horizontal or vertical, which describes the height or width of an object. This unit is measured along the pencil or pen using the thumb. Once this initial line is made on paper, successive lines can be drawn in proportion to it, as they are measured using the pencil.

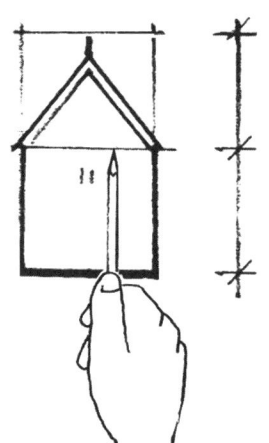
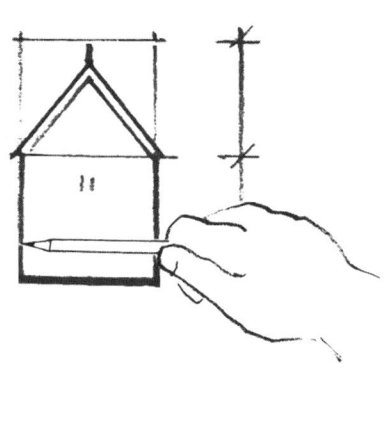

Construction methods
From the plan view
Using a perspective grid
From photos
Drawing freehand perspectives
Estimating proportions
Freehand one-point perspective
Freehand two-point perspectives
Atmospheric perspective
Graphic emphasis

6

The units measured with the pencil are useful construction guidelines with which to estimate the correct proportions of elements to one another. The length demarcated on the pencil does not have to be transferred exactly onto the paper surface.

The size of the drawn object is one of the variables in freehand drawing. How long a line is drawn or the size on object on paper is completely up to the sketcher. This method only helps to achieve the right proportions.

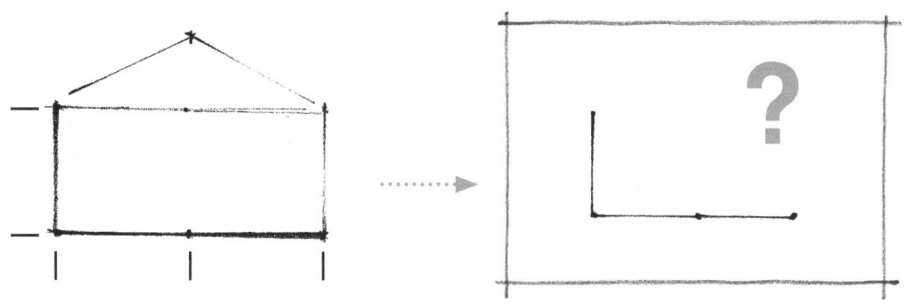
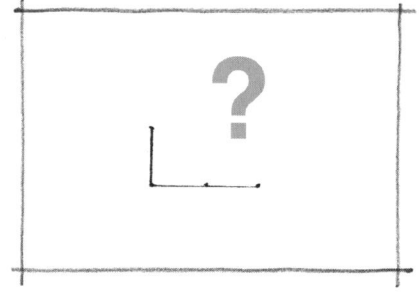

In this façade, the measured height to width ratio is 1:2.

This proportion is then drawn onto the paper surface....

...but it is up to the author of the drawing to decide how large it will be shown.

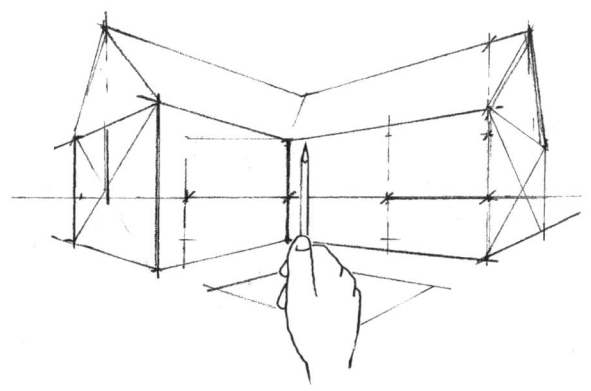
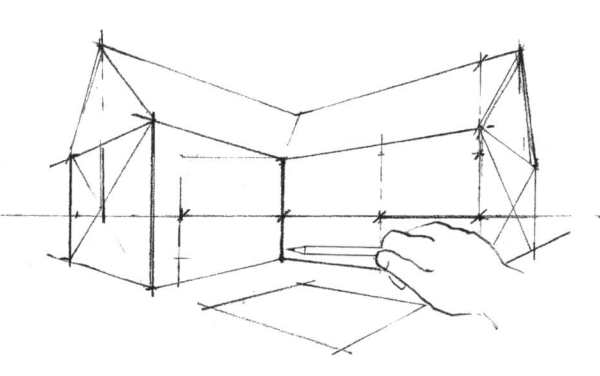

For a two-point perspective without a frontal elevation, the measuring line will be a vertical corner or edge. This height can be estimated using the pencil and all other horizontal proportions are measured as if the seen object is on a flat surface.

Don't hesitate to include construction lines when building the perspective scene. These help to ensure correct relationships of all elements to one another and can easily be erased later.

Perspective

Freehand one-point perspective

The principles of linear perspective are still applied when we are outside and drawing a scene with one clear vanishing point. The task might prove a little more difficult, as we are in the scene we are drawing. The horizon line and the upright picture plane are invisible to us, but they must be kept in mind as we construct our drawing.

Every plane and vertical edge parallel to the picture plane remains parallel. All other parallel lines converge at the vanishing point determined by our very own eye level.

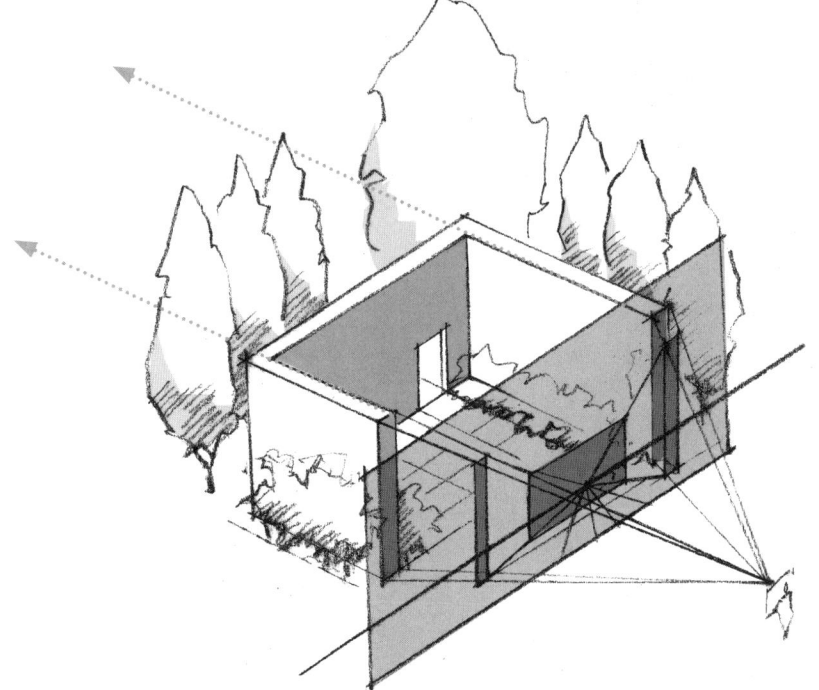

It is helpful to visualise this perspective set up when drawing outdoors. It is also useful to hold up the drawing surface as if it is the picture plane upon which the scene emerges. What is parallel and what is not can easily be determined when seen in the context of the vertical drawing surface or picture plane.

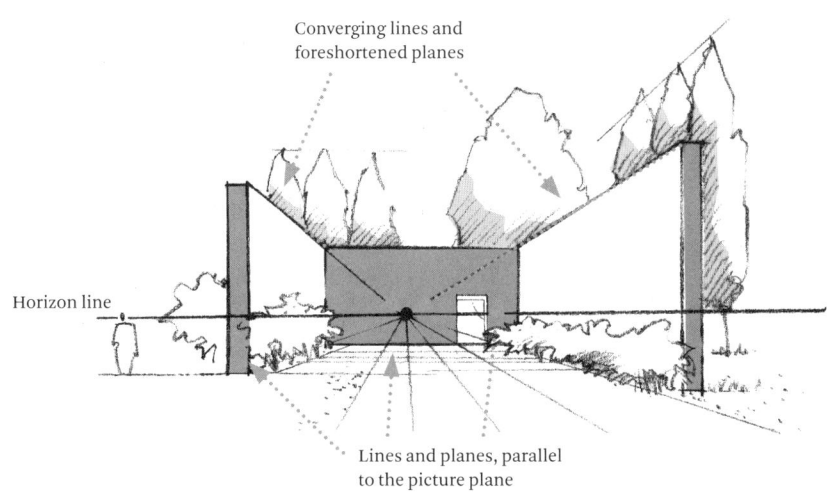

Converging lines and foreshortened planes

Horizon line

Lines and planes, parallel to the picture plane

Construction methods
From the plan view
Using a perspective grid
From photos
Drawing freehand perspectives
Estimating proportions
Freehand one-point perspective
Freehand two-point perspectives
Atmospheric perspective
Graphic emphasis

6

The principles found in perspective construction are clearly seen in this sketch. All the square boxwood hedges have a horizontal edge which lies parallel to the picture plane. The other edge converges towards the vanishing point.

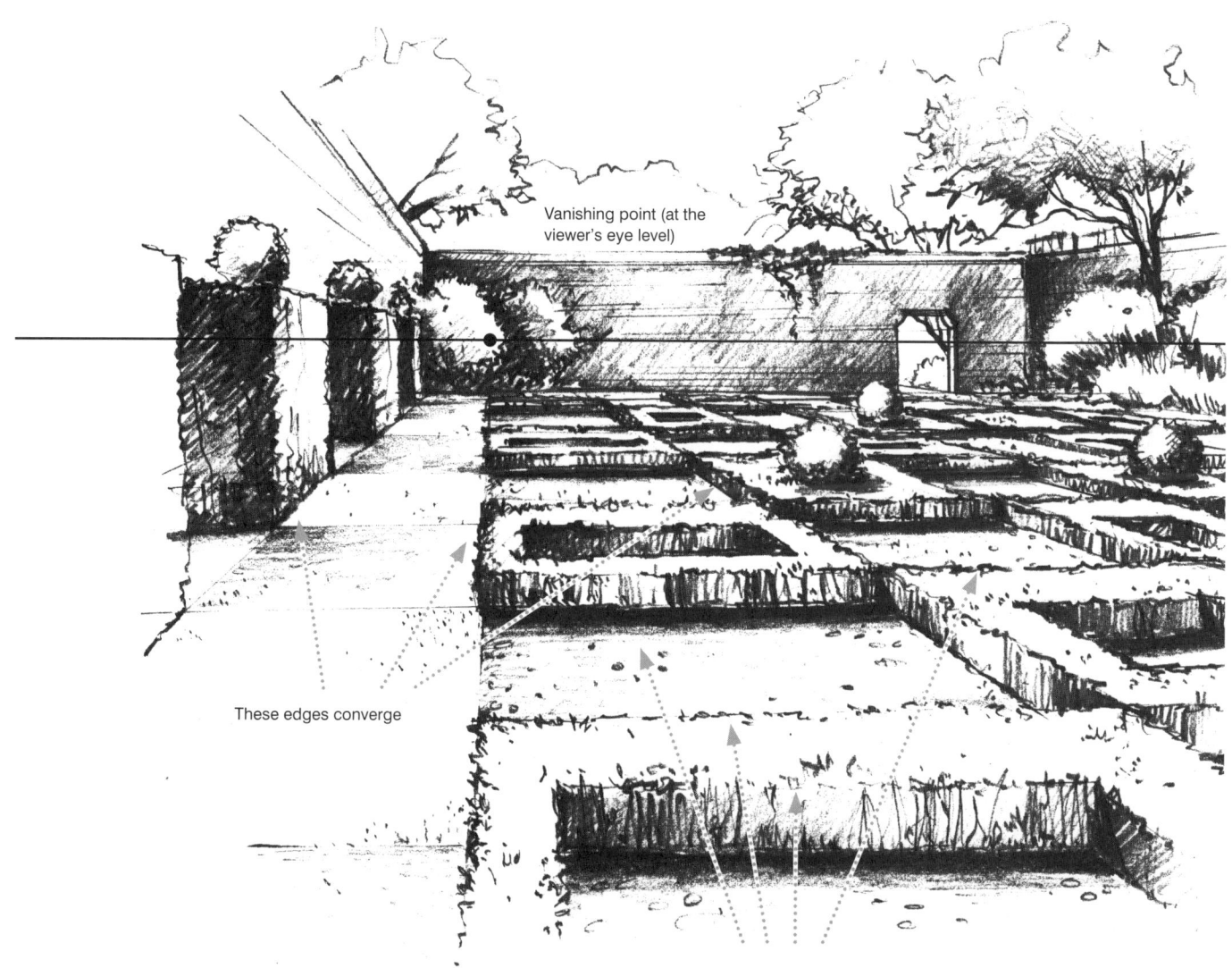

Vanishing point (at the viewer's eye level)

These edges converge

These horizontal edges lie parallel to the picture plane. They don't converge. Instead, they get smaller as they get closer to the vanishing point.

Perspective

Freehand two-point perspective

It is more common to draw two-point perspectives on site, since they appear less static and more natural to us. Again, it is useful to keep the perspective set up in mind when drawing a two-point perspective scene.

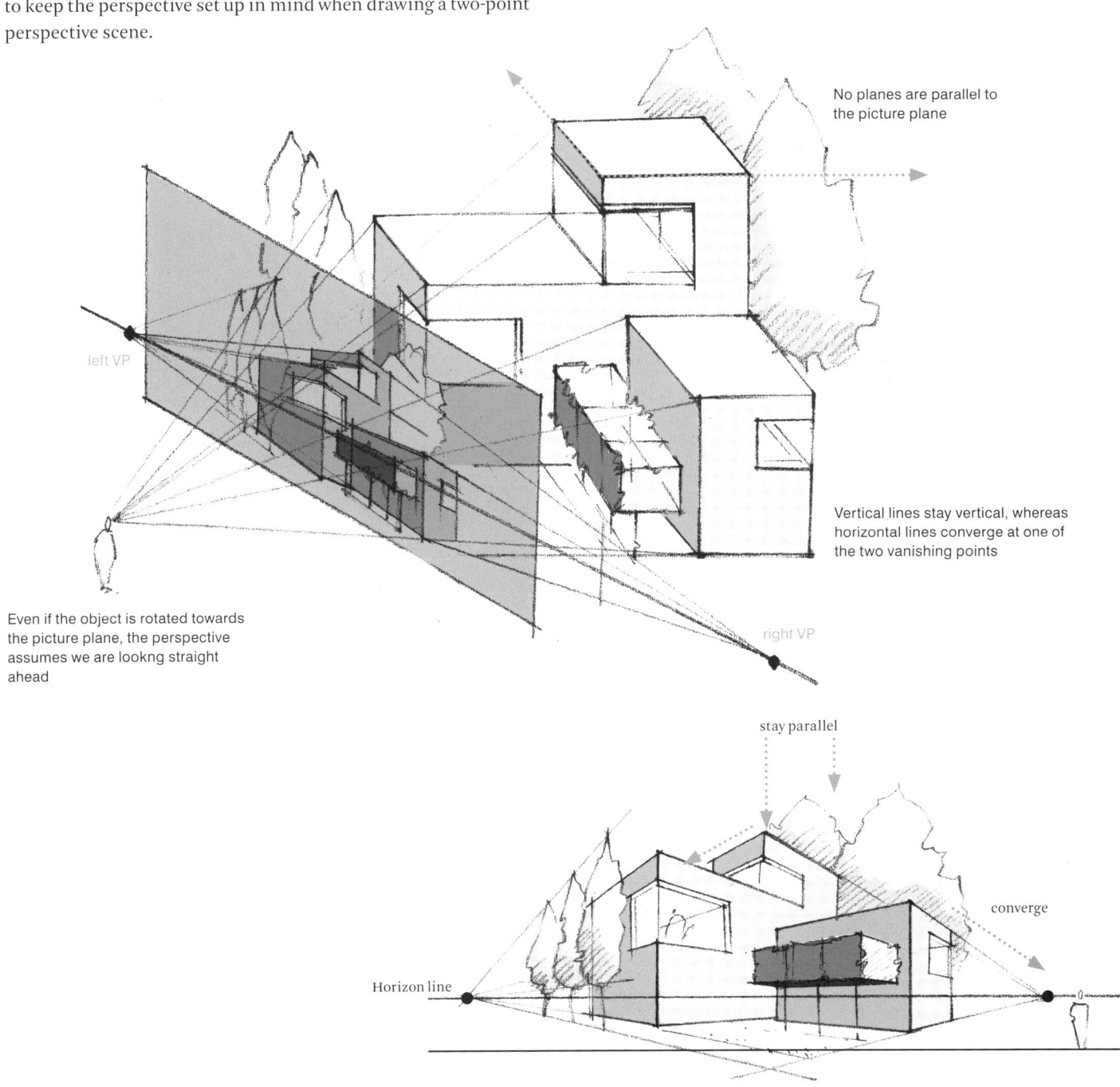

No planes are parallel to the picture plane

left VP

Vertical lines stay vertical, whereas horizontal lines converge at one of the two vanishing points

Even if the object is rotated towards the picture plane, the perspective assumes we are lookng straight ahead

right VP

stay parallel

converge

Horizon line

Construction methods
From the plan view
Using a perspective grid
From photos
Drawing freehand perspectives
Estimating proportions
Freehand one-point perspective
Freehand two-point perspectives
Atmospheric perspective
Graphic emphasis

6

Since landscape architects don't always only draw buildings, the two-point perspective view of a space is particularly useful. When looking into a space which is rotated to our picture plane, visualising the vanishing points right can be tricky. The right planes converge towards the left and the left planes converge towards the right.

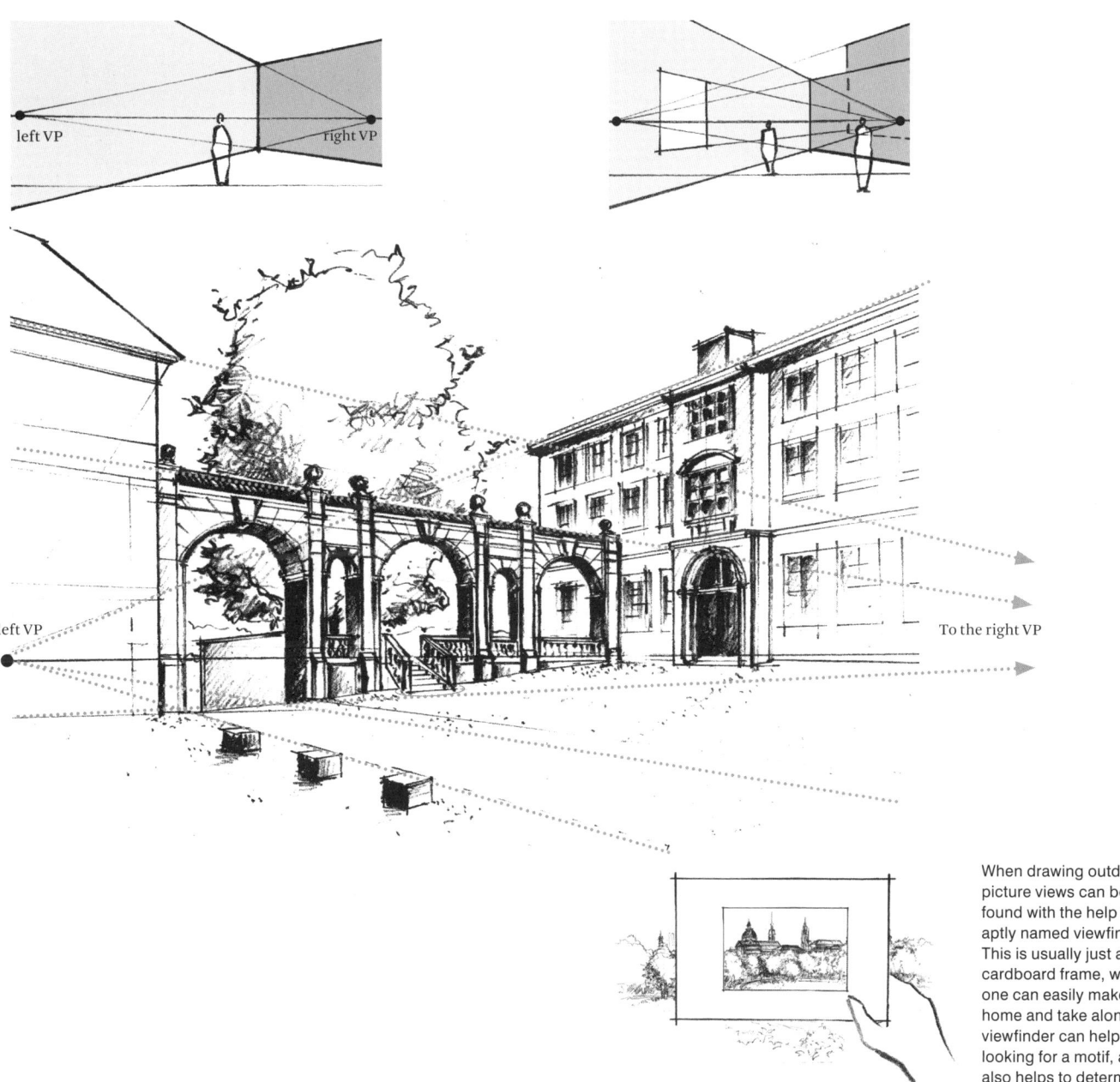

When drawing outdoors, picture views can be found with the help of an aptly named viewfinder. This is usually just a small cardboard frame, which one can easily make at home and take along. The viewfinder can help when looking for a motif, and it also helps to determine the borders of a drawing.

Perspective

The atmospheric perspective
Freehand perspective drawing not only relies on linear perspective construction, it also frequently borrows principles found in atmospheric perspective. This type of perspective set up takes its cues from landscape painting. Rural or even wild landscape situations are not usually constructed, that is to say they do not have intersecting orthogonal lines. This means there are often no vanishing points to be found in a scene. An atmospheric perspective, sometimes also called aerial perspective, achieves depth by layering elements and progressively muting their tonal values and contrasts across a pictorial space. The illusion of distance from foreground to background is achieved by overlapping elements and by reducing the clarity, intensity, size and contrast with which objects are shown.

The picture plane becomes a layered stage set. Elements distributed and arranged at the bottom serve as the foreground. Elements at the top of the picture plane are perceived as being further away and become the background. This moves our eyes upwards from the bottom of the picture plan, up to the top, from front to back.

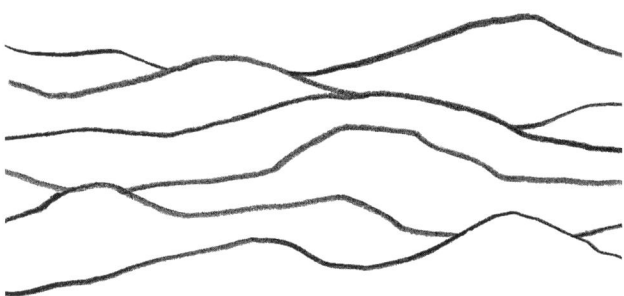

Overlapping and layered elements

This so-called depth cue seems to pull some elements forward and push other elements back, depending upon their tonal values and contrasts. Muting tones and the sharpness with which elements are shown in the upper part of the picture plan will visually push them into the background.

Varying contrasting light and dark areas from foreground to background give a pictorial effect of depth

Construction methods
From the plan view
Using a perspective grid
From photos
Drawing freehand perspectives
Estimating proportions
Freehand one-point perspective
Freehand two-point perspectives
Atmospheric perspective
Graphic emphasis

6

Varying object sizes can suggest distance. The different sizes of the elevations shown on the right immediately convey a sense of depth. Larger objects are automatically read by viewers as being closer.

Depth is suggested by varying object sizes

Increasing contrast in larger and nearer objects also brings them forward in a scene. By reducing the contrast of the background elements, they appear to be less visible and seem further away. This so-called atmospheric haze is common feature when we look into a distant landscape. The areas furthest from us seem to become lighter and less clear as they fade into the distance.

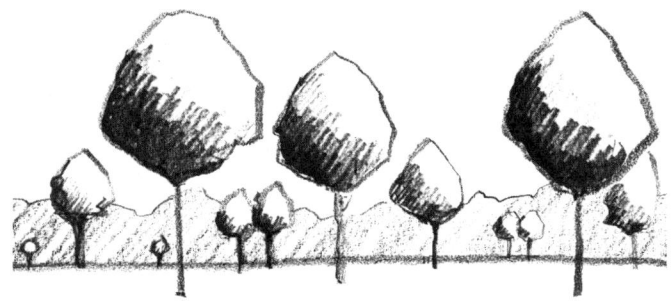

Variations in contrast from foreground to background

Another trick for achieving a sense of depth is to vary the detail in the objects shown. This means that foreground elements and objects can be very detailed, whereas the smaller background elements are left with little more than their outlines.

The further away an object is meant to appear, the less dark, detailed and sharp it needs to be.

Differing degrees of detail from foreground to background

Perspective

Rendering freehand perspectives: Light and shade

At the end of this chapter, it must be mentioned that a good freehand perspective does not end with its correct construction. Along with good composition on the page, a graphically convincing perspective requires rendering – with textures, tonal values, light and shade – to finally emerge as a stimulating scene. These techniques, which cannot be further elaborated upon here, not only add depth but also graphic emphasis to the perspective. The author of the drawing can distinctly highlight his or her intentions when finishing a scene, bringing objects of interest to the forefront of the drawing and letting less important elements fade away. There are infinite ways with which to render a scene and to add interest. The simple examples below demonstrate how the use of contrast, tonal values, light and shade can affect the way a perspective drawing looks. It is a good idea to try out different rendering variations, in order to achieve the best possible graphic and pictorial effects in the finished picture.

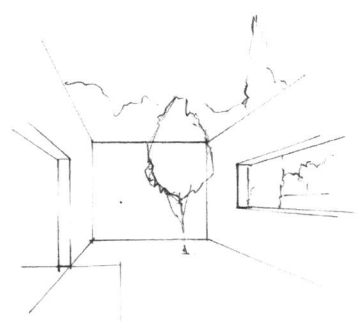

Construction methods
From the plan view
Using a perspective grid
From photos
Drawing freehand perspectives
Estimating proportions
Freehand one-point perspective
Freehand two-point perspectives
Atmospheric perspective
Graphic emphasis

6

Architectural presentations, layout and lettering

Layout
- 220 Introduction
- 222 Formats
- 223 ISO Standards (DIN)
- 224 Symmetry
- 226 Asymmetry
- 230 Montage
- 233 Ordering information

Adding words to a presentation
- 234 Text size and hierarchy
- 235 Key words vs. the legend
- 236 Hand lettering
- 238 Futura alphabet

Architectural presentations, layout, and lettering

Layout

To layout a presentation means to order all of its parts into a legible and effective composition. Viewers must engage with it quickly and follow the hierarchy of presented information easily. They should be able to understand the designer's key ideas, the design process and developments which led to the final design scheme. The goal of every good layout is communication. The macro-layout deals with overall layout systems and structures, geometric structures and formats, which underlie any good presentation. Micro-layout concerns itself with the hierarchy and order of information, typography and text, spacing and margins, as well as the distribution of graphic weight and emphases on a sheet, screen or board. A layout binds together all parts of a presentation and must be considered carefully if it is to transmit the important graphic and text components of a project.

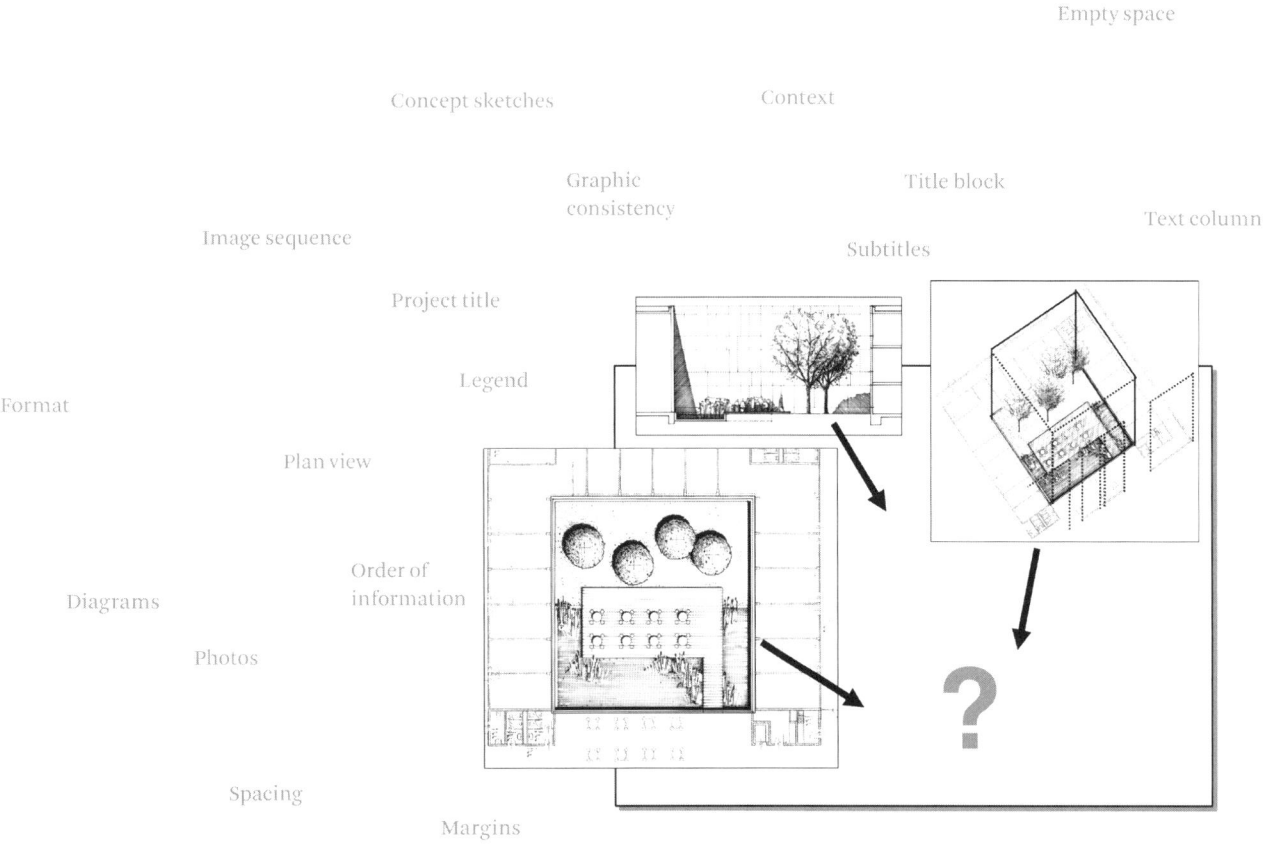

Layout
Introduction
Formats
ISO Standards (DIN)
Symmetry
Asymmetry
Montage
Ordering information

7

A landscape architectural presentation will never be composed of images alone. It combines graphic information with text, and often is comprised of several sheets of interrelated information, each corresponding with the overall project. An effective layout should result a harmonious and easy legible overall composition, which communicates the design idea and all of its components to its viewers, regardless if they are laypersons or other professionals. The following pages in this chapter can only briefly discuss this very extensive topic. There are countless good books on the subject of layout, and these should be referred to for more detail on this important subject.

Graphic information

Plan view / Floorplan
Elevations
Section / Section-elevation
Axonometric / Isometric
Perspectives
Diagrams
Background spaces
Images / Photos.....

Text and annotations

Title / Subtitle
Body of text
Legend
Title block
North arrow & graphic scales
Dimensions & measurements
All other necessary information relating to the project that cannot be drawn....

Architectural presentations, layout, and lettering

Standard paper sizes

When deciding on a layout, one of the first questions to be asked is usually about the presentation format. Will the sheets have portrait or landscape orientation? A successful layout depends on the way in which all parts come together. No one part of the composition should distract from another, but rather the arrangement of the many drawings and contents should reinforce each element. The result is a holistic overall composition, which is much more interesting that the individual drawings on their own. Elements should form a graphic unity without appearing dull. All parts must be interrelated, successively communicating the important aspects of a design project in a continuous and logical sequence.

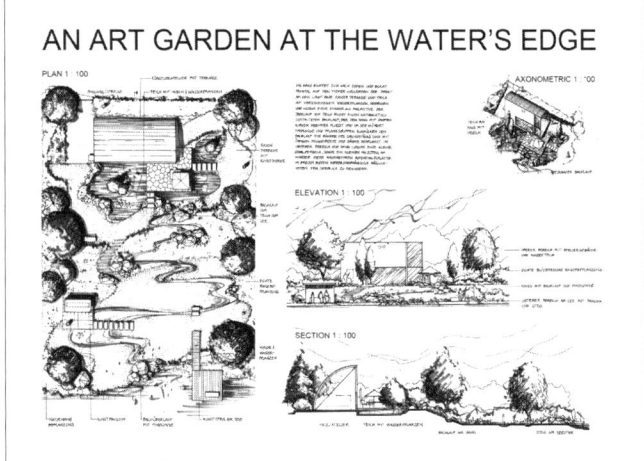

For landscape architectural presentations, the appropriately named landscape format is more common...

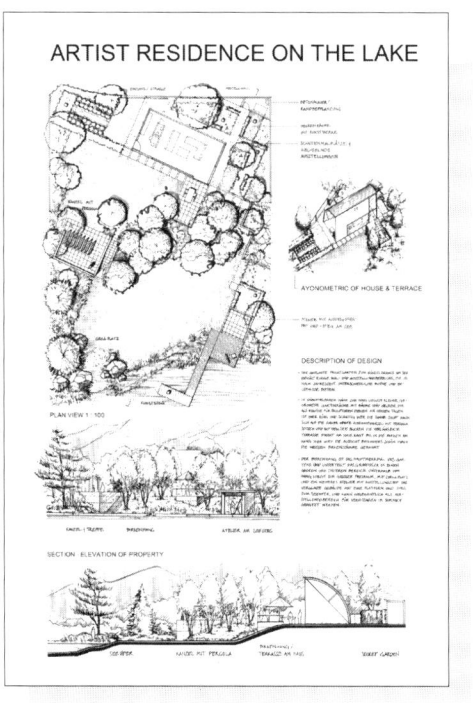

...however, the portrait format is equally effective. It has a more graphic quality, much like a poster.

Layout
Introduction
Formats
ISO Standards (DIN)
Symmetry
Asymmetry
Montage
Ordering information

7

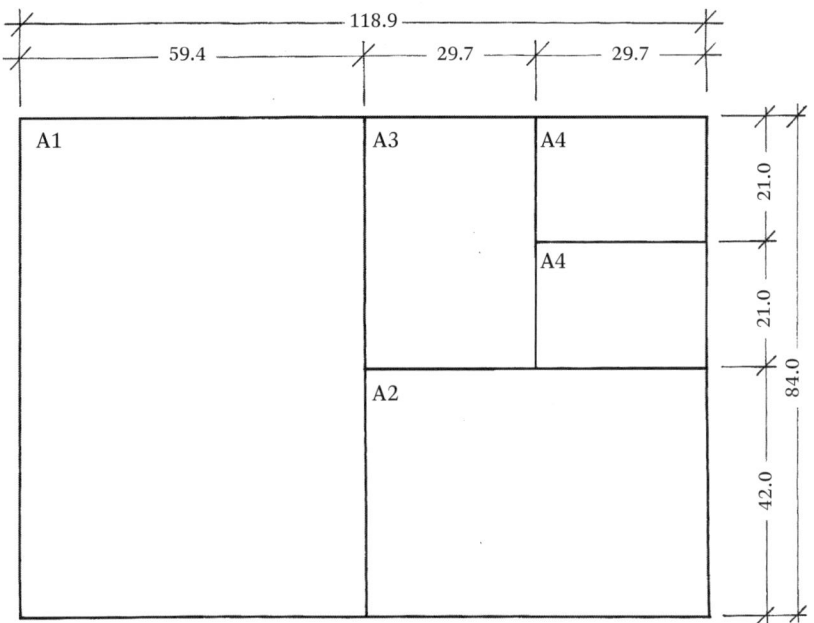

International paper size standards are sometimes referred to as ISO standards. These follow the German DIN 476, whose proportions are based on a ratio of 1 : 1.41.

The A-series are frequently used, especially for larger presentation boards. These are labeled as A1, A2, A3 and so on.

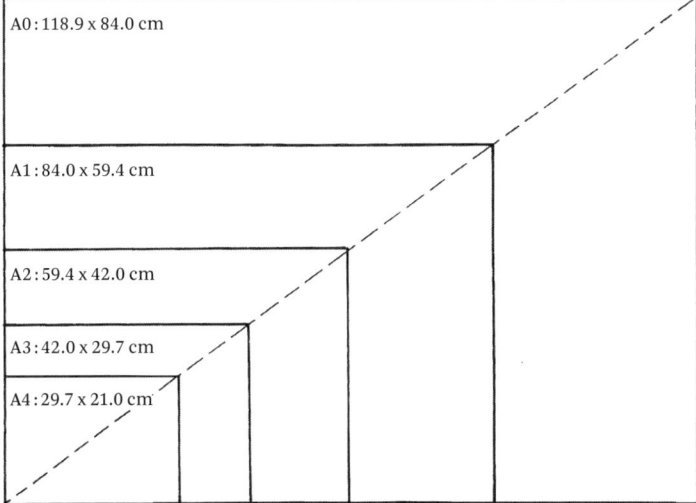

Each successive paper size is defined by halving the previous size along the larger dimension. The proportions remain constant throughout all formats.

Architectural presentations, layout, and lettering

Symmetrical layouts
A symmetrical layout is organised and subdivided using a central axis, either vertical or horizontal. This line, whether visible or not, structures all content in a presentation.

The contents are positioned and distributed along or in distinct relationship with this main axis.

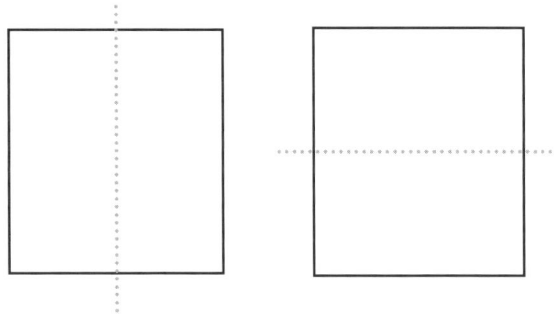

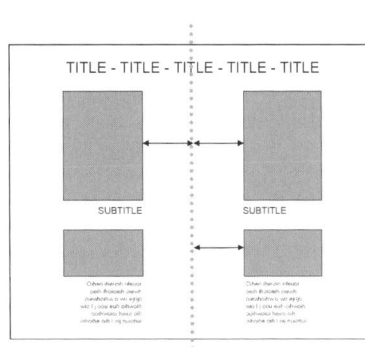

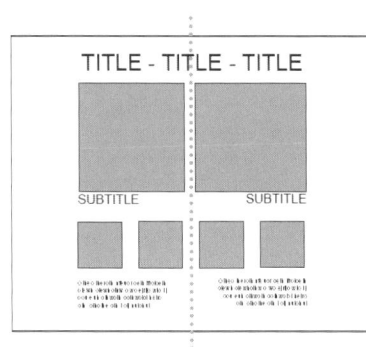

Spaces and gaps between images are important, whether along an axis or not. The closer two images come together, the more unified they appear.

The vertical axis orders the spacing and placement of all elements adjacent to it. All parts are positioned equidistant to the middle line.

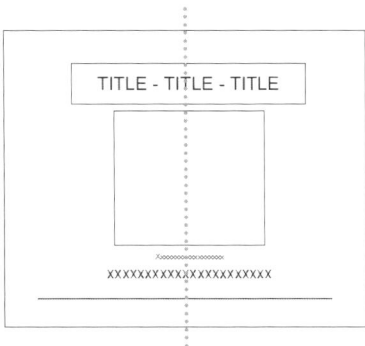

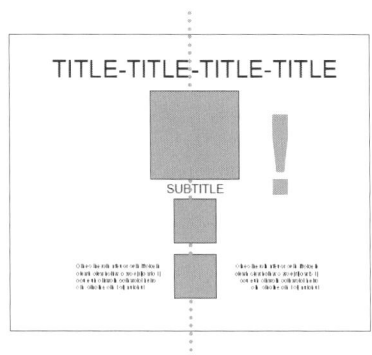

Vertical axes work best with vertical formats. Too much left over space can quickly appear empty.

Layout
Introduction
Formats
ISO Standards (DIN)
Symmetry
Asymmetry
Montage
Ordering information

7

Symmetrical presentations characteristically appear calm and balanced. The symmetrical organisation must be upheld as much as possible in order not to disturb this effect.

The graphic weight of the individual elements in a presentation (their darkness, contrasts and tonal values) must be considered and organised effectively in order to achieve to achieve the desired balance.

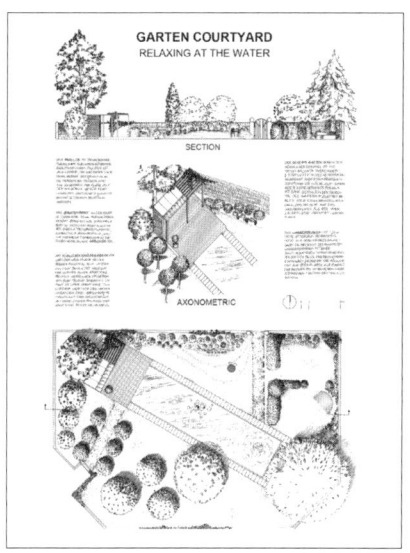
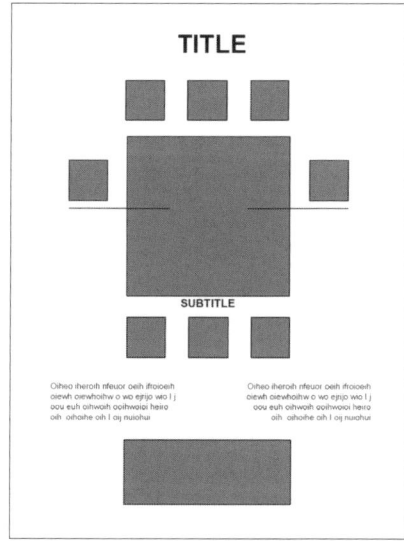
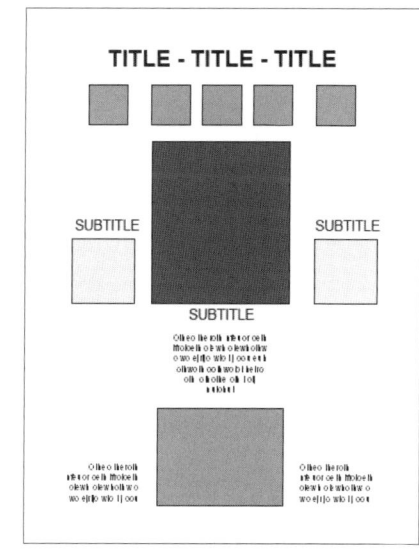

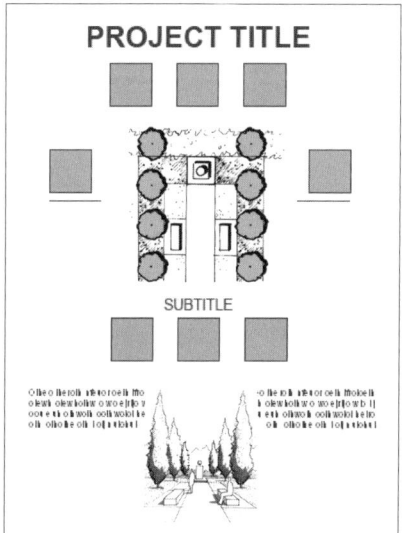

Symmetrical layouts work best with symmetrical design projects

225

Architectural presentations, layout, and lettering

Asymmetrical layouts

Asymmetrical layouts are far more common in landscape architectural presentations. Although they are composed of many different-sized parts, these layouts must still achieve a visual balance and easy legibility. Compared to the more static and calm symmetrical layouts, asymmetrical layouts have increased potential to create graphic tension between their different-sized individual parts. The asymmetrical composition is usually the result of different scaled drawings which communicate all aspects of a design. In landscape architecture, the main – and often the biggest – element in a presentation is the the site plan. This generally takes on the most graphic weight in a presentation. It is usually surrounded by smaller elements and text which enhance and inform the plan. Examples of such layouts are shown on the following pages and are meant to give an idea of the many possibilities to consider when laying out a presentation.

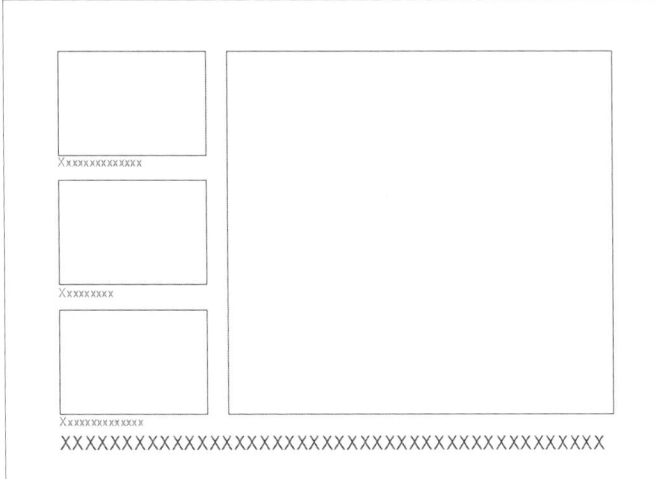

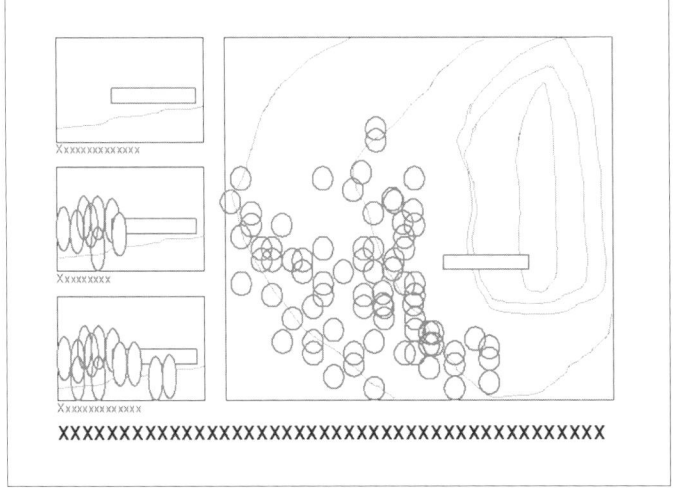

Smaller drawings, photos and text information can be positioned around a central plan in order to highlight its contents

Layout
Introduction
Formats
ISO Standards (DIN)
Symmetry
Asymmetry
Montage
Ordering information

7

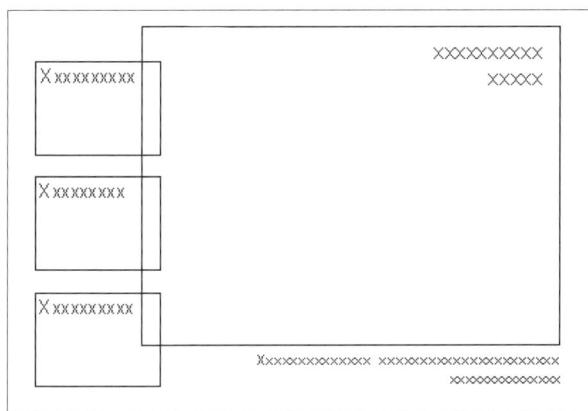

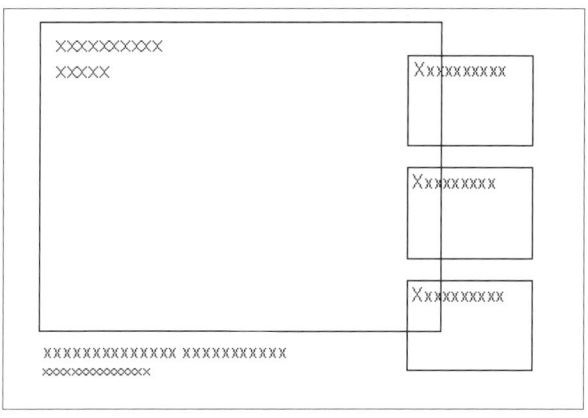

Individual images can overlap each other or even be positioned within a site plan, as long as all important information remains visible

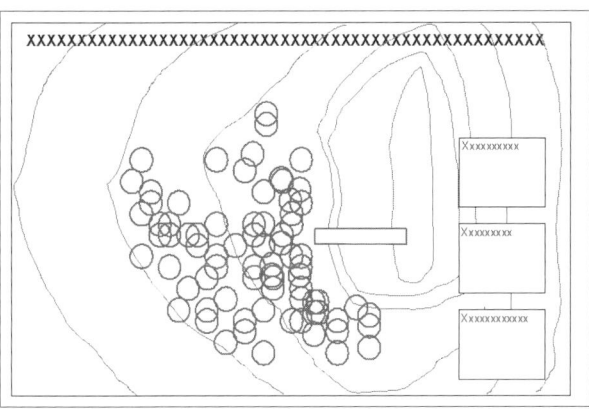

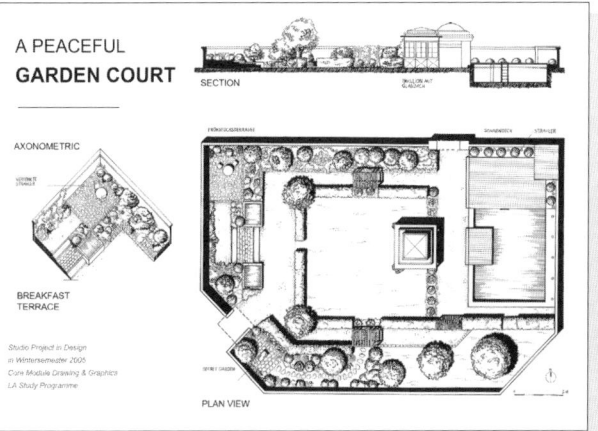

Elements within a presentation do not necessarily need a frame or box outline. However, these can sometimes help to unify many different drawings on a sheet.

Other orthographic projections which inform the plan are usually aligned with it. The relationship between the projections must be clearly legible and ensure that the overall composition is visually balanced.

Architectural presentations, layout, and lettering

Dividing up a presentation sheet into three equal parts, either vertically or horizontally, is a common and visually pleasing layout method

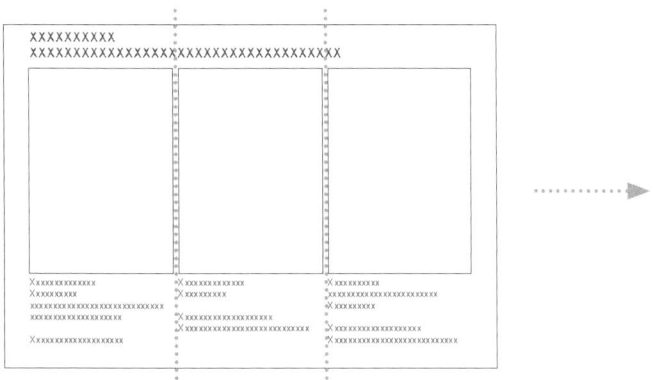
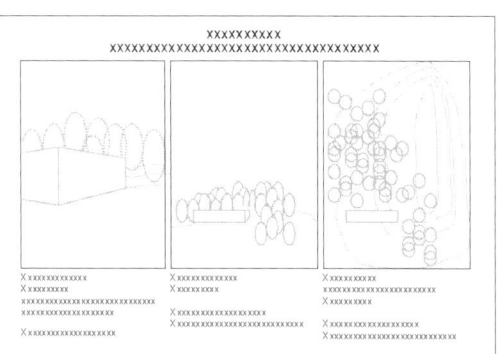

The tri-partite subdivisions can organise both graphic and text contents

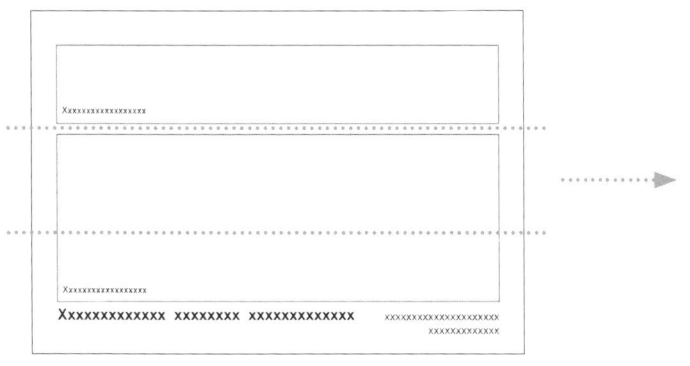
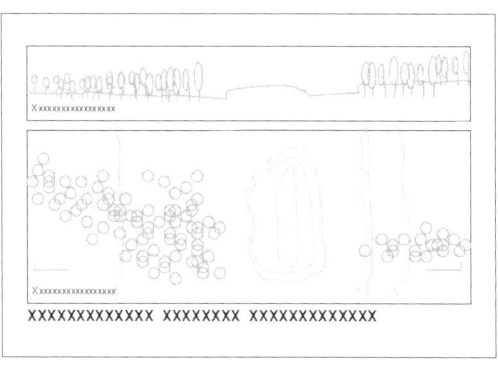

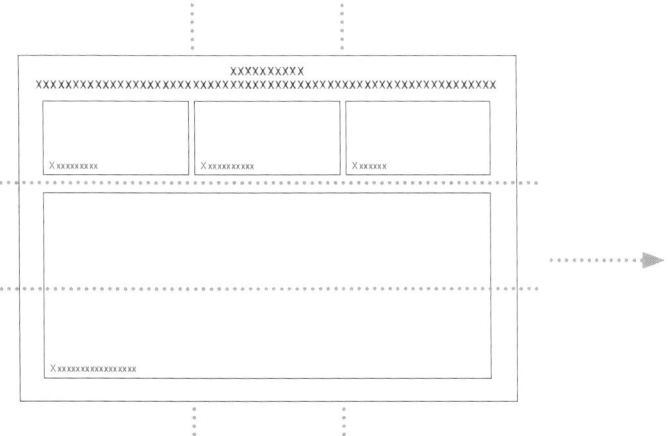
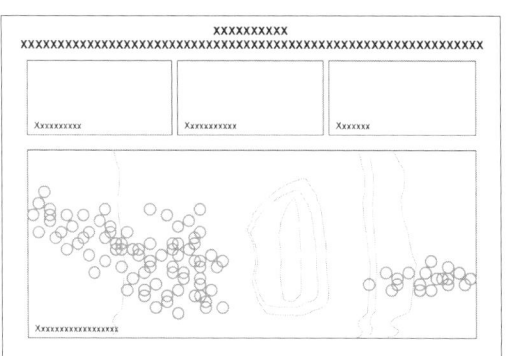

Layout
Introduction
Formats
ISO Standards (DIN)
Symmetry
Asymmetry
Montage
Ordering information

7

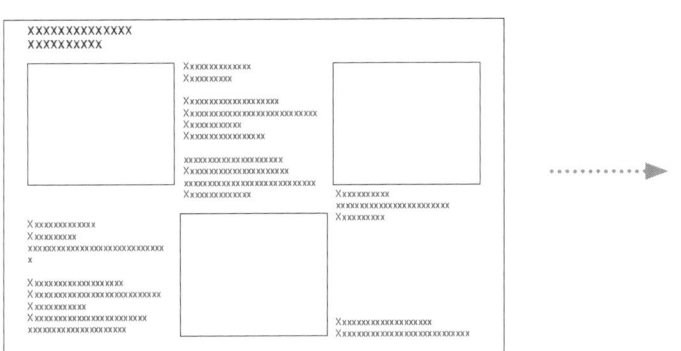
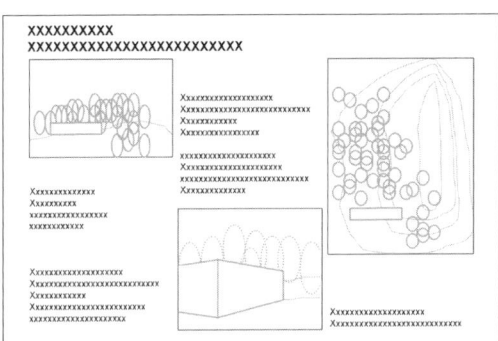

The layout is the underlying framework which governs all parts of a presentation. Regardless of the geometric subdivisions or structures chosen, a good layout will to successfully bring together all of the very different contents, even down to spacing and margins.

The more irregular the elements within the layout, the more important it is to clearly establish the relationships between them. A good presentation should always look balanced and be easily legible.

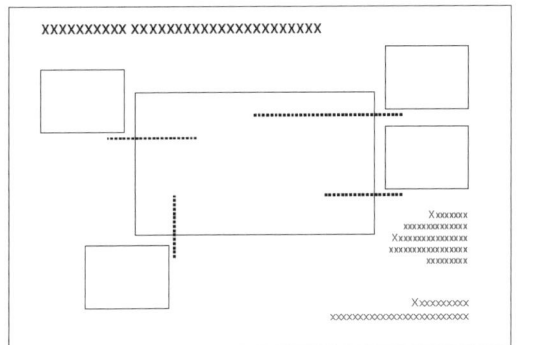
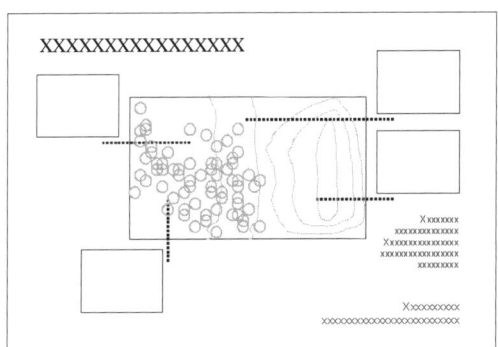

Architectural presentations, layout, and lettering

Montage: Putting the elements together
There is never a single right way to put together a presentation. It is helpful to layout the elements in different variations and look at them from a distance before deciding upon and finalising the chosen final structure.

Whether done by hand or with the help of a digital program, a layout will have the same goal: to appear as a continuous, legible and harmonious sum of all parts.

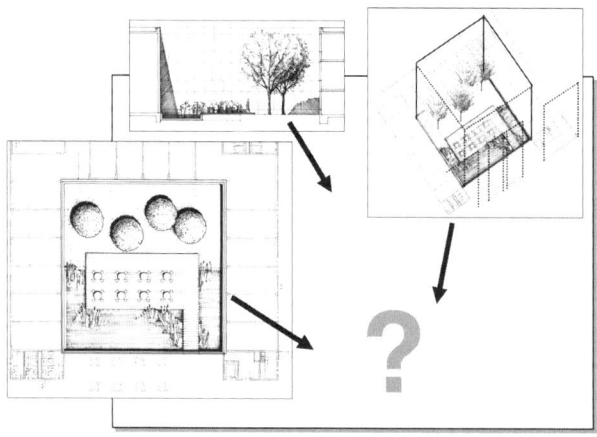

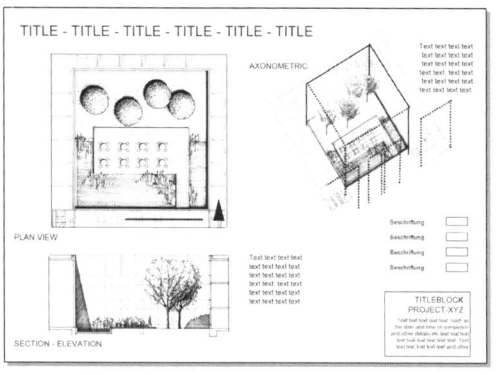

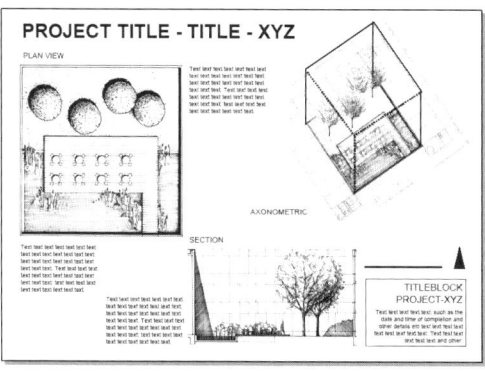

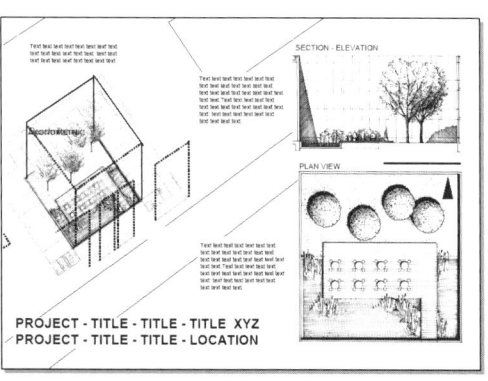

Presentation drawings are usually related. Aligning the orthographic projections helps to read the information together, as shown in the examples on the left.

Layout
Introduction
Formats
ISO Standards (DIN)
Symmetry
Asymmetry
Montage
Ordering information

7

Empty spaces in the presentation plan can be infilled with textures, colours or even photographic images. Make sure to use these elements as background information, so that they do not distract from important content.

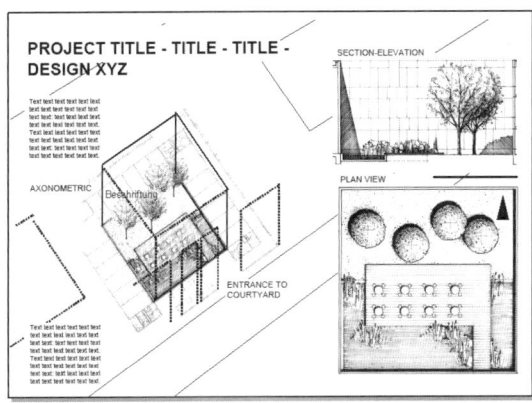

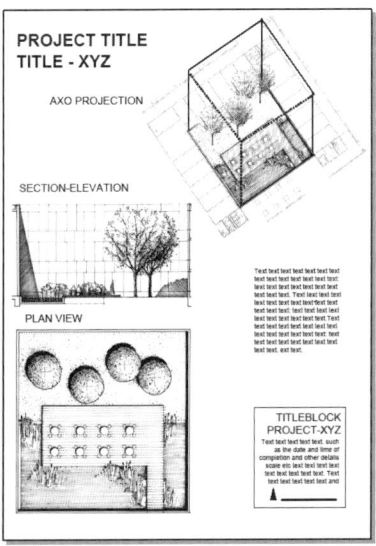

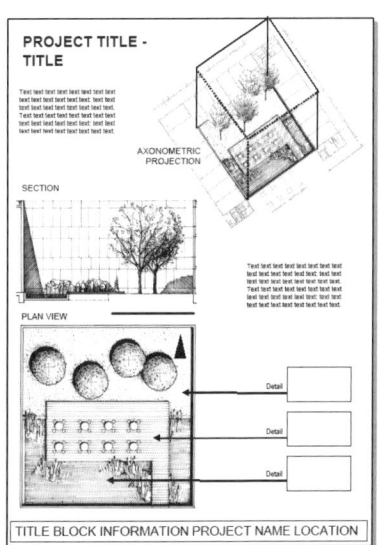

If unsure about whether a layout is finished, it helps to hang up the sheet, step back from it and look at it from a distance. Turning it upside down is another helpful method for determining if the composition is balanced.

It should quickly become clear whether or not and important information is communicated effectively.

Architectural presentations, layout, and lettering

Other layout tips

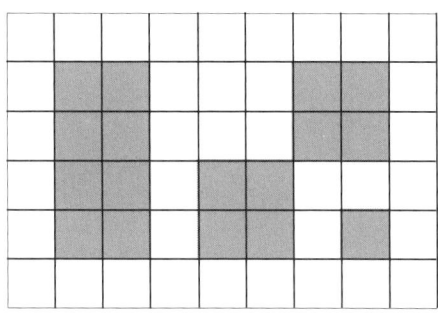

Some designers find it helpful to use an underlying grid to layout the presentation sheet and to organise all the individual elements. Using a grid is also a good way to bind different elements together, especially when presentations contain more than one sheet.

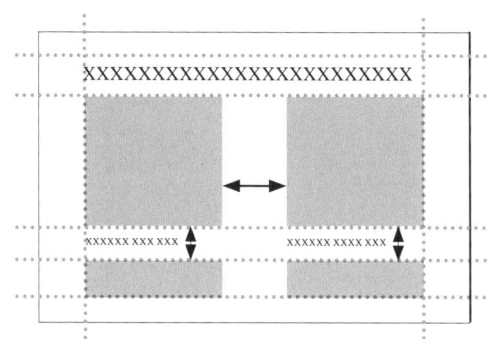

When positioning pictures and text, always ensure they are reasonably and sensibly spaced from one another. If drawings and text appear too close, they begin to form a visual unity or perhaps even compete with each other, becoming difficult to read. The more contents on a presentation sheet, the more blank space has to be introduced to allow the eye to scan and uptake individual portions of information. The size of these areas will depend on the graphic weight of the individual drawings.

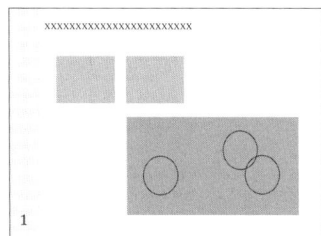
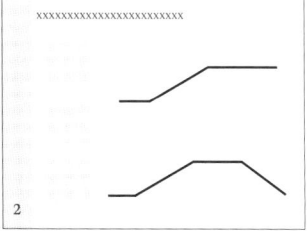
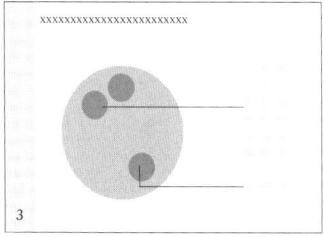

If there are several sheets in a single presentation, they should share an overall layout structure which repeats itself. This repetition binds the various contents together, even if they are very different.

Recurring colours, graphic elements and style, as well as a consistent typography, will help to unify different sheets within a presentation

Layout
Introduction
Formats
ISO Standards (DIN)
Symmetry
Asymmetry
Montage
Ordering information

7

Sequence of information

The shapes, sizes and tonal values of a drawing define its graphic weight in a composition. Larger, darker and more contrast-rich drawings will automatically draw more attention and must be positioned carefully. They will always appear to carry more graphic weight.

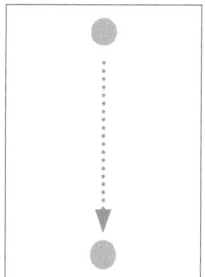 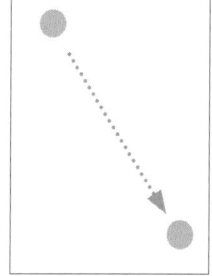 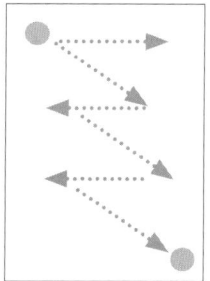

Architectural presentations in the Western world tend to be read from left to right and from top to bottom, whether on paper or in digital form. The viewer must be clearly guided through the layout and receive continual portions of information in the correct sequence. It is a good idea to start with an overview of a design project and lead the viewer from the overall context down to the smaller details in a logical way. Although the individual parts may be drawn eloquently, if the layout is not put together effectively, the information or, worse, the viewer's interest in the project, may be lost.

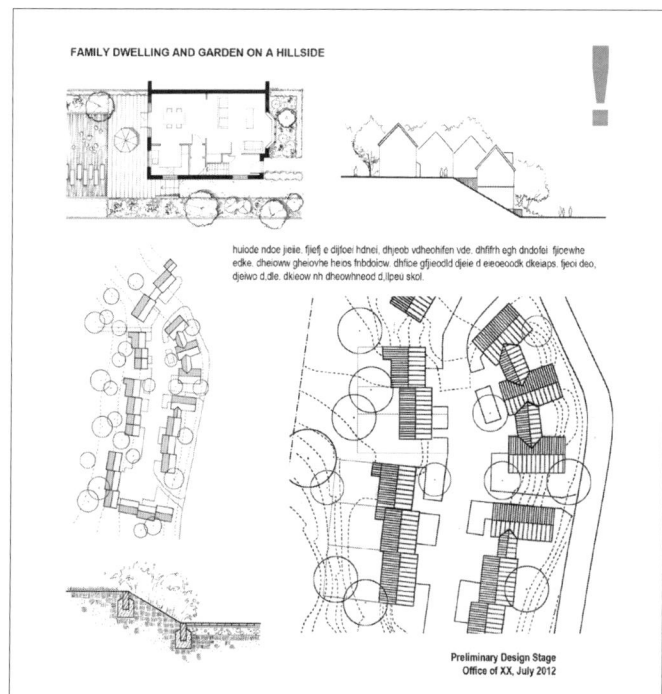

This presentation does not appear logical because the emphasis is given to the wrong plan view

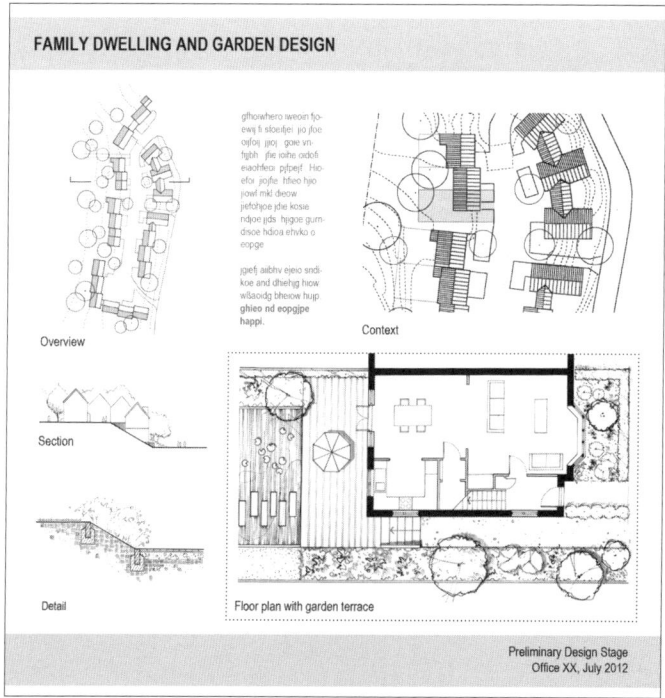

The emphasis here is clearly on the garden design, which occupies the most space and thus carries the most graphic weight in the composition

233

Architectural presentations, layout, and lettering

Adding text to a presentation
Text and lettering in a presentation must be legible from different distances. Text size must be composed within a hierarchy.

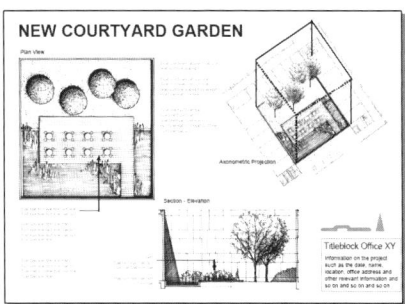

The most important information is biggest, as this text is read first. It is usually the title of the project, written in upper case letters, instantly informing the viewer of the project. The secondary information comes next, which may be subtitles or key words. Then comes all other text information, to be read from very close range. These may be smaller blocks of text and the title block at the bottom right of the plan.

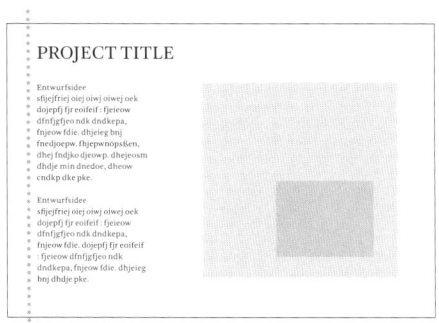

Blocks of text should be presented in columns, much like they would in a newspaper or magazine

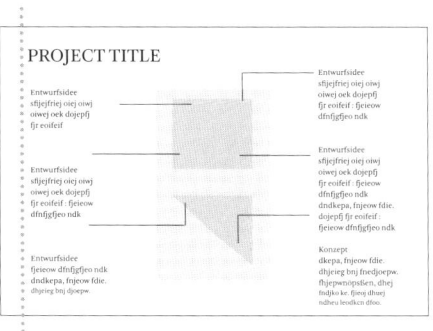

If text is positioned outside of drawings, align and relate it clearly to the drawing it describes and always justify the edges. Use thin, orthogonal arrows to help connect text and image.

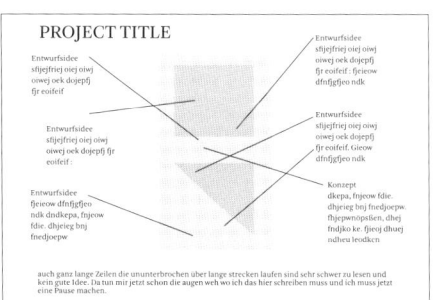

Too many arrows with different directions can disturb the composition. Arrows should never cross each other. Non-justified and unevenly placed text blocks make a sloppy and unprofessional impression.

Long sentences running across the wide expanse of a sheet are uncomfortable to read. Keep text in small and compact sets, which add to the overall visual composition.

Adding words to a presentation
Text size and hierarchy
Key words vs. the legend
Hand lettering
Futura alphabet

7

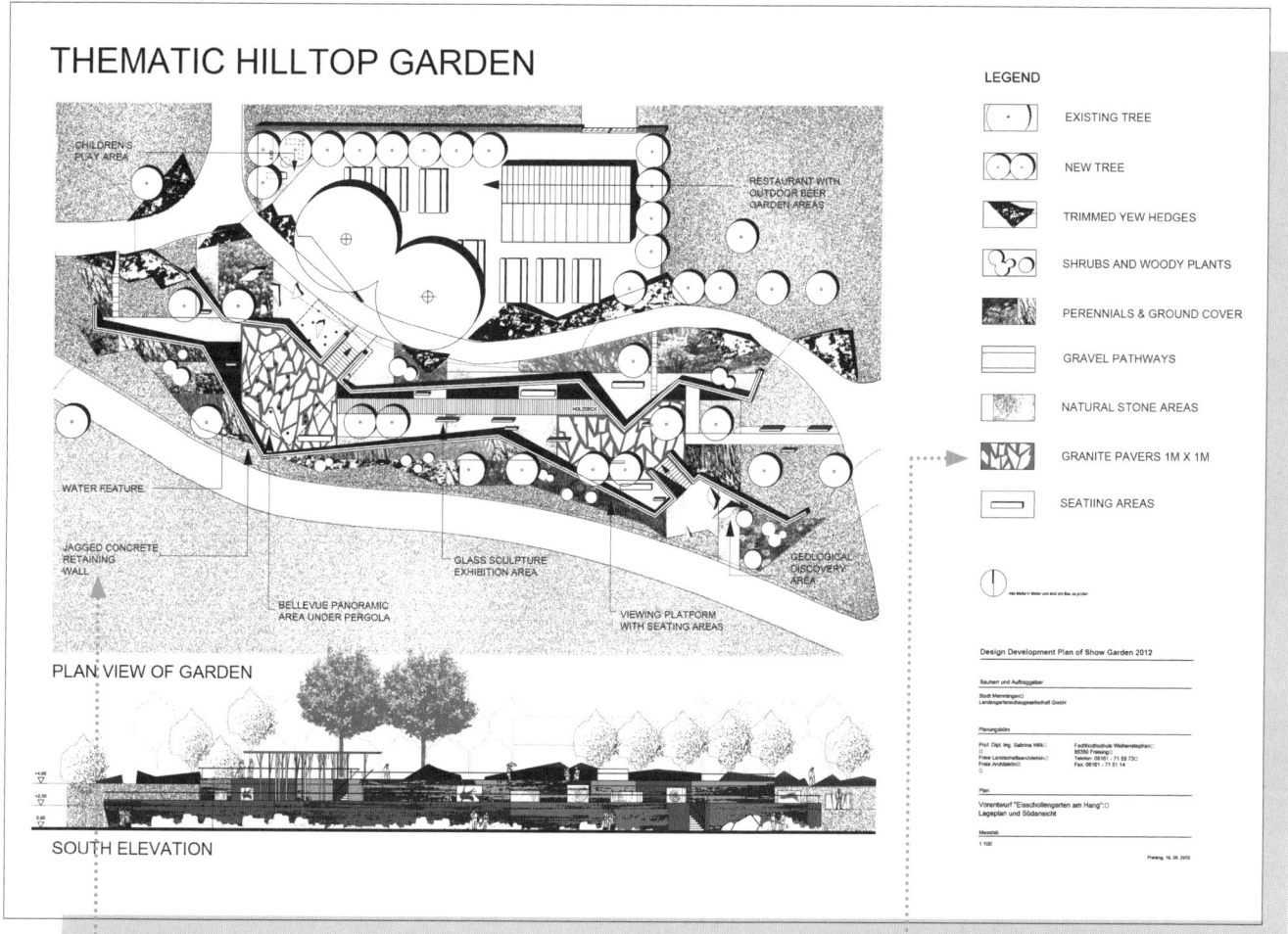

Key words in and around the plan can help the viewer to locate uses, buildings and functional areas. They allow for quick orientation before all of the graphic components are fully read.

The legend, which is sometimes also called a key, is essential to organise the different surfaces, materials and textures occurring in a plan.

The legend contains small squares or rectangles, each filled with excerpts of textures and colours relating to contents seen in the plan. It can also contain symbols. The legend forms a type of chart which is usually located directly alongside the plan it informs.

Architectural presentations, layout, and lettering

Hand lettering
Even though these days most presentations are in electronic format, there are still situations where good hand lettering is required. A computer may not always be at hand on a construction site or during a design discussion meeting. Clean and clearly legible lettering will make a good impression.

For hand written annotations it is easiest to print words using only capital letters without any of the serifs usually found in handwriting. They are a good accompaniment to rectilinear sketches and can be learnt quickly.

ABCDEFGHIJKLMNOPQRSTUVWXYZ

Regardless of their size, it is important to keep letters consistently upright. This gives lettering a uniform character. If necessary, use guidelines, both horizontal and vertical, to ensure consistency.

ABCDEFGHIJKLMNOPQRSTUVWXYZ

1 2 3 4 5 6 7 8 9 10

Lower case letters have a less formal effect and may also look livelier than capital letters

abcdefghijklmnopqrstuvwxyz

Adding words to a presentation
Text size and hierarchy
Key words vs. the legend
Hand lettering
Futura alphabet

7

At the beginning it will feel strange to apply lettering to sketches by hand, simply because we are so used to clicking and typing. With practise, we will also get a feel for the right spacing and consistency.

Inevitably we all develop our own individual style of lettering. Lettering should always enhance a drawing, be easy to read and pleasing to look at. Irregular and uneven lettering can distract from a drawing, making the overall presentation look sloppy.

The larger the lettering, the more dominant it will appear in a presentation. Larger letters will also need stronger line weights otherwise they will appear too weak.

ABCDEFGHIJKLMNOP
QRSTUVWXYZ

Practising lettering is easily done on graph paper and with the help of guidelines. A T-square and a triangle can help to achieve good geometric proportions, however hand lettering is usually done freehand and not only at the drawing board.

ABCDEFGHIJKLMM
MNOPQRSTUVWXYZ

Spacing between lines of text will also need to increase with the lettering size

237

Architectural presentations, layout, and lettering

Futura alphabet
One of the most common lettering types is Futura.
These letters are a good starting point for practising as they are based on a square and proportional subdivisions of it.

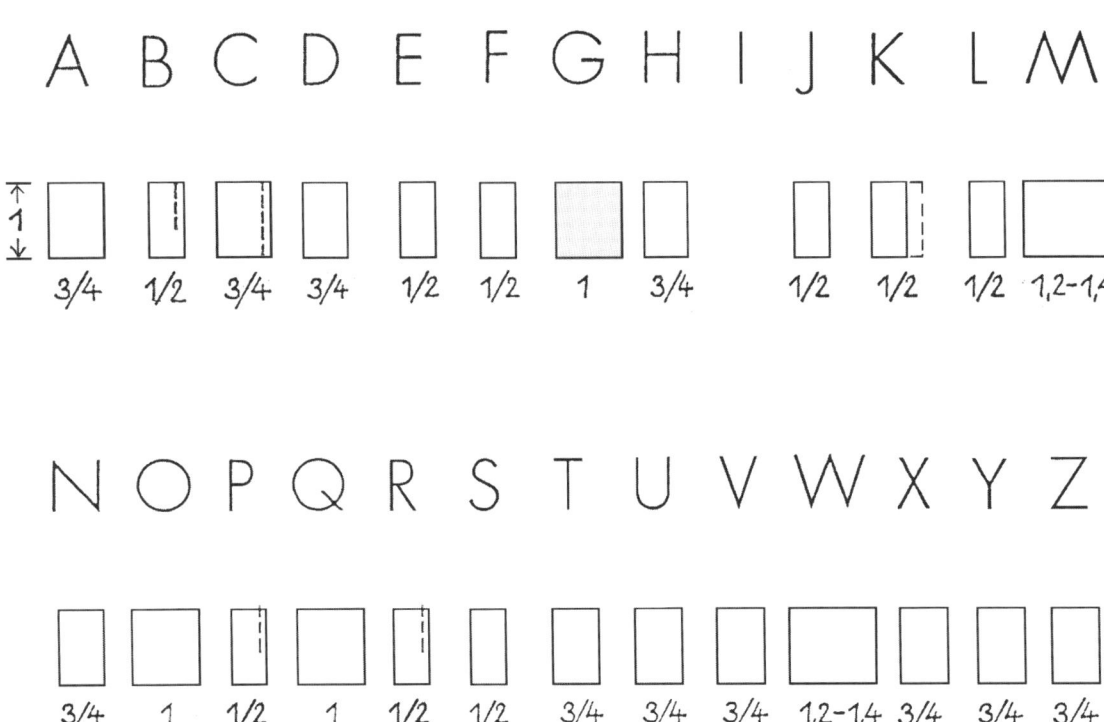

Adding words to a presentation
Text size and hierarchy
Key words vs. the legend
Hand lettering
Futura alphabet

7

The topic of typography and text is an exciting one, and can only be briefly mentioned here. There are countless books on typography, fonts and layout and I can only recommend that landscape architects engage with this important topic.

Although we are not trained as graphic designers, we need to learn the impact of typography on the layout and realise that it must be carefully selected and considered as part of any presentation. Do not forget, we do not only read text, we see it as well. It has as much of a graphic effect as any drawing.

The examples shown here are lettering exercises. Each text uses the Futura alphabet in three different sizes, varying their compositions on the page.

239

Appendix

242 **Final thoughts**
243 **Acknowledgements**
244 **Index**
245 **Bibliography**
246 **Picture credits**

Final thoughts

Drawing is an enjoyable activity. It can be done almost anywhere, is inexpensive and fun. The best sketches are often spontaneous and quick, without any laborious effort or deep thought. The best time to learn how to draw and how to use these skills effectively at university. It is also the best time to develop one's own graphic handwriting.

This book is merely a starting point and only scratches the surface of this vast topic. The reality is that drawing in landscape architecture is much more complex and compelling than what is presented in the pages of this book. We have only discussed basic line drawing, coupled with drawing conventions and methods which have been taught and practised as part of studies of landscape architecture and architecture for centuries. Our use of drawing is usually for a specific purpose as we are generally bound to communicate and visualise existing and new spaces for others graphically. Our dreams and visions, regardless of whether in plan view, elevation, section or perspective, must always engage with an audience. Ultimately, they serve to get our ideas transformed into real, built environment. But drawing in landscape architecture also embraces observation, perception, exploration, analysis, colour theory, design principles, freehand drawing and collage, not to mention technical draughtsmanship and construction drawing at different scales. Due to constraints of space, these topics are not covered here.

Many of my students believe that drawing ends after their university course. Looking at today's workplace, we might be inclined to think that this is true. However, the reality is very different. Drawing is still a part of professional practice in many different disciplines, including landscape architecture. Naturally, it comprises a large part of the artistic and creative professions, where the image is paramount. Professionals in graphic design, film, art, illustration, architecture and building, as long as those working in automotive, industrial and fashion design, all still use drawings and sketches to develop, explore, visualise and present their ideas for things that will, one day, become real.

There are also less obvious professions that require sketches and drawings in their day-to-day work. Their goal, is not to produce a lovely picture, however, but rather to express an idea or to make the invisible visible. Doctors sketch surgical procedures, physicists document experimental processes, choreographers sketch movement sequences and stage sets, statisticians formulate results, engineers and mathematicians draw machines and formulas. Drawing is done in almost every field.

It is much more than achieving a pleasurable end result. A sketch can communicate and document, can clarify complex thought processes, can even distil and display hidden emotions. Psychotherapists, programmers, journalists, archaeologists, geographers, musicians and scientists are just a few of the professionals for whom drawing plays an important role. Regardless of mathematical formula, technical machine or trendy T-shirt, most objects created begin with a simple sketch. The same idea applies to landscape design, which is also a man-made construct influenced by many different factors.

I often hear my students claim that they just cannot draw. This is simply not true; each one of us can draw. It does take discipline and practice to become adept at drawing and, unfortunately, in our fast-paced digital age, we have become accustomed to being able to achieve instant results. So cast your aspirations to be Michelangelo aside for a while and just start making marks, observing and sketching the world around you. And do it at every opportunity. Try out as many styles and effects in your sketchbook as you can and you will be amazed at how quickly you progress. The only way to learn to draw is by drawing.

Drawing and sketching give you the freedom to communicate and express your unique ideas and visions. Once the fundamentals are grasped, they are never forgotten. And, if nothing else, drawing is an enjoyable and relaxing task. It remains a creative and very immediate personal experience, an activity which is the opposite of clicking and staring at a computer screen. Drawing is a wonderful counterpart to digital media and remains an important part of thinking, visualising and communicating in landscape architectural design.

One final point remains to be made. Communication is an increasingly important topic in a globalised world. Even after 2000 years of application, drawing still offers exciting new possibilities alongside today's CAD-drawings and computer-apps. Just as speaking many different languages expands your world, the same principle applies to drawing. It allows you to graphically communicate with many different people from different backgrounds, to express yourself, to clarify ideas, and even to delight and inspire others. I hope this handbook helps as a first step in understanding the possibilities within landscape architectural drawing and inspires readers to explore their own ideas creatively on paper, both during and beyond the drawing classes in landscape architectural studies.

Acknowledgements

This handbook is intended to be both instructional and inspirational. The inspirational part would not have been possible without the help and support of many individuals. I would like to thank and acknowledge all the contributors from various countries in both professional practice and academic institutions for their time and generous efforts to make this publication a reality. During the writing of this book, I had the pleasure of both getting in touch with old colleagues and of establishing new international contacts as well.

I will start with many thanks to Professor Maria Auböck (Academy of Fine Arts, Munich) and her team at Auböck-Karasz landscape architects in Vienna for taking the time and effort to contribute her sketches, inspite of a very busy schedule. My thanks to Katrin Albrecht and Britt Blom, along with partners Peter Schatz, Wolfgang Betz, Michael Kaschke and Henrike Wehberg-Krafft, Claus Rödding and Hinnerk Wehberg from WES Landscape in Hamburg for putting together and allowing me to show so many beautiful drawings in this book. Thanks to Kamel Louafi and his team in Berlin for allowing me to include his wonderful freehand sketches.

I am indebted to Prof. Dr Udo Weilacher for his help with the acquisition of Prof. Dr Dieter Kienast's drawings and for assisting with the introductory text for Kienast. I thank Dr Anette Freytag from the ETH in Zurich, who went to great lengths to source the drawings I was looking for and to contact Mrs Erika Kienast for approval. Anette is currently finishing a comprehensive publication on Dieter Kienast's life and work. I thank Alex Winiger at the ETH archive for his help in locating Kienast's drawings and also to Marion Brakebusch at Vogt Landscape Architects for searches on drawing locations. I owe Mrs Kienast a considerable debt of gratitude for allowing me to show her husband's exquisite work in this book.

Many thanks go to Prof. Chip Sullivan at the University of California at Berkeley for his extraordinary drawings, diaramas and cartoons. His books were a great inspiration to me as a student. Thanks too to Prof. Marc Treib, also of the University of California at Berkeley for his help and advice on Garrett Eckbo's drawings and to Miranda Hambro at the university's Environmental Design Archives for her assistance.

I am extremely grateful to my talented local Bavarian colleagues for all of their support with this book. I thank Harry Dobrzanski (TU Munich/ Die Grille, Penzberg) for his incredible drawings and collage-montages, Axel Lohrer and Ursula Hochrein (lohrer.hochrein landschaftsarchitekten, Munich) for their wonderful sketches and Dr Monika Supé for allowing me to publish her sketches and layout examples. Thanks to my colleagues at the University, Justus Thyroff and Wolfdietrich Rahm for the drawing exercises, Matthias Thoma for the perspective grids and especially Ingrid Schegk for her enjoyable support with my drawing excursions. A very special thanks goes to my assistant for drawing and design Thomas Grubert (Grubert architects, Penzberg) not only for his eye-catching drawings and travel sketches but also for his tireless support, enthusiasm and energy he has put (and continues to put) into my drawing and design courses at the university.

Big thanks go to my colleagues in the United Kingdom. I thank Christoph Brintrup, Bernie Foulkes and Peter Corrie at LDA Design in London for the enthusiastic support of this book and for contributing so many wonderful drawings. Thanks go also to Ina Woeste for her assistance and advice with drawings in the book. I am also grateful to Robert Rummey and Elizabeth Staveley from Rummey Design in Sevenoaks and in particular to Jane Brightwell for putting together so many wonderful sketches. I thank Petra Funk, Kathryn Gustafson, Neil Porter and Mary Bowman from the London office of Gustafson Porter for their spectacular presentation drawings of Terrasson and Westergasfabrik. Thanks to Winnie Poon at West 8 in Rotterdam for locating the drawings. Thanks go out to Belinda Chan, Derek Lee and the team at PWL Partnership in Vancouver, Canada for putting together so many incredibly eye-catching and persuasive drawings for me – your time and efforts are much appreciated.

The second edition of this book includes the wonderful sketches and drawings of atelier le balto, Berlin. My thanks go to Marc Pouzol for his generosity. I am also very grateful for the addition of Sabine Heine's exquisite and inspirational illustrations to this new edition.

I would also like to thank and acknowledge my students and alumni whose work is featured in this book. The fact that I get to work with such enthusiastic students is a great inspiration to me. I am always grateful to be a part of their academic careers and to be able to play a role in the development of their graphic abilities. It is a pleasure to support and advise so many talented young people. A very big thanks to Prof. Dr Wolfgang Sonne at the TU Dortmund and Prof. Dr Kathleen James-Chakraborty at University College Dublin, who encouraged me to publish this book. And, finally, heartfelt thanks to my publisher Philipp Meuser and his team at DOM publishers, in particular to graphic designer Daniela Donadei, Masako Tomokiyo and Stefanie Villgratter for their patience and support in making this book. It has been a rewarding experience.

Index

Aerial perspective 187
Architect's scale (triangle scale) 14
Arbour 82, 134
Asymmmetric layout 226
Atmospheric perspective 214
Axonometric projection from plan (plan oblique) 34, 161
Axonometric projection from elevation (elevation oblique) 34, 158

Boulders 91
Building sections 142
Brick 136
Bicycle stands 97
Bird's eye view 214
Blossoms 72

Circles 162, 192
Circle template 14, 46, 57
Central projection 34
Construction 21, 39, 82, 110, 154, 161, 172, 183, 190, 194, 201, 209, 216
Construction drawings 39
Converging lines 175, 177
Co-ordinate systems 182
Cone of vision (perspective) 175, 183
Contour lines 98
Contrast 16, 22, 41, 54, 71, 74, 77, 102, 104, 107, 114, 123, 148, 165
Cut surfaces (section) 147

Depth cues 121, 123, 128, 132, 167, 177, 182, 214
Desk area 15
Diagonal lines (perspective) 185, 190
DIN-Norms 222
Distance (perspective) 183
Drawing surface (plane) 21, 33, 110, 175
Drawing styles 7, 119
Dry stone walls 91, 136

Effects (graphic) 6, 12, 16., 22, 24, 26, 29, 32, 35, 43, 53, 61, 64, 76, 82, 106, 114, 119, 121, 128, 132, 136, 159, 165, 167, 177, 182, 186, 201, 203, 214, 222, 226, 239
Effects (spatial) 32, 83, 121, 215
Elevation 110–114

Elevation oblique 34, 160
Eraser 10
Expression (graphic) 26

Flexible ruler 14
Floor plans 40
Flowers 72
Foreshortening 177
Fountains 92, 138
Freehand drawing 10, 88, 172, 206–217
Furnishings 40, 96–97
Futura alphabet (typeface) 238

Glass (elevation) 137
Grid (perspective) 184, 198
Grid, in layout 232
Ground cover 70, 130
Ground plane (perspective) 176
Grasses 64, 130
Gravel surface 87

Hedges 16, 64, 115, 180, 211
Horizon line 177, 186
Human figures 96, 139

Ink pens 11, 16
Isometric projections 34, 159

Key (legend) 235
Key words (presentation) 235

Layout 220
Layout, symmetrical 224
Layout, asymmetrical 226
Leaf patterns 44, 74, 118
Legend 235
Letters (typeface / alphabet) 236
Lettering 234
Linear perspective 144, 174, 177–178 190, 194, 206, 210, 214
Lines 16
Line weights 22
Line hierarchy 22

Markers 12-13, 18–19, 41, 55, 57, 69
Marble surface 137
Materials 26, 60, 86, 136, 146, 167, 235
Walls 87, 90, 101, 136, 142, 210

Natural stone surfaces 87, 136
North arrows 102, 221

Observation (sketching) 80
Orthographic projections 32
One-point perspective 178, 180,
Outlines (shapes/ forms) 28, 41, 44, 56, 64, 105, 111, 114, 116, 202
Overlapping 177

Paper (formats) 15, 222
Parallel projections 34, 158
Parallel ruler 15
Paving patterns & built surfaces 86, 136
Picture plane 21, 32, 175, 183, 193, 200, 210
Pencils 10, 16
People 96, 139
Pergola 82, 134
Perennials 72, 130
Perspective 172
Photos (perspective overlay) 204
Pigment pens (ink pens) 11
Plan oblique (axonometric project.) 34, 161
Planting beds 76
Plasticity 12, 41, 44, 47, 49, 53, 56, 106, 114, 118, 128, 132, 16, 193, 215
Potted plants 74
Preliminary sketch 38, 104
Presentation 25, 43, 204, 220, 222, 224, 226, 229, 232, 236, 239
Projections 32,158
Proportions (sketching) 208

Ramps 90, 147, 188, 201
Reflections (perspective) 189
Rendering tools 13, 18
Retaining walls 101
Rocks 91
Roof plans 40
Roof shapes 41

Scales (graphic) 102
Section 140
Section arrows 144
Section elevation 145
Sequence of information 233
Shade & shadows 41, 47, 53, 57, 61, 76, 82, 90, 105, 118, 164 177, 193, 216

Bibliography

Shrubs 64, 128
Site plan 21, 38, 42, 46, 53, 92, 94, 98, 101, 105, 110, 115, 138, 147, 194, 220, 226
Size diminution (perspective) 177
Sketchbook 15, 80
Sketch paper 15, 204
Spot elevations 100, 167
Stairs 90, 147, 188, 201
Standard formats 222
Stones 91
Surfaces 11, 18, 26, 86, 98, 106, 136, 177

Terrain 99
Three-point perspective 34, 179
Topography 98
Trace paper 15, 204
Tree symbols 42, 116, 163
Tree groups 56, 122
Triangle 14
Tripartite layout 228
T-Square 14
Two-point perspective 34, 178, 212, 180

Unity (layout & presentation) 103, 220, 230

Vanishing point 34, 173, 177, 181, 188, 196, 205, 207, 212, 214, 215
Vegetation 70, 80
Vellum (drawing paper) 11, 15
Volumes 70, 131

Water 92–93, 138, 189
Windows (elevation) 128, 137
Winter trees 58
Wood 10, 86, 113, 137
Woody plants 66, 128, 164
Work space 15

Amoroso, Nadia:
Representing Landscapes:
A Visual collection of landscape and architectural drawings,
Abingdon, Oxon 2012.

Andrews, Jonathan:
Handgezeichnete Visionen,
Berlin 2004.

Ching, Francis:
Drawing – A Creative Process,
New York 1990.

Ching, Francis:
Design Drawing,
New York 1997.

Ching, Frank:
Architectural Graphics,
New York 1996.

Dee, Catherine:
Form and Fabric in Landscape Architecture,
Abingdon 2001.

Dines, Nicholas:
Landscape Perspective Drawing,
New York 1990.

Elam, Kimberley:
The Geometry of Design,
New York 2001.

Elam, Kimberley:
Typographic Systems,
New York 2007.

Enthwistle, Trudi / Knighton, Edwin:
Visual Communications for Landscape Architecture,
London 2013.

Gänshirt, Christian:
Tools for Ideas,
Basel 2007.

Itten, Johannes:
The Art of Colour,
Ravensburg 1961.

Lin, Mike:
Architectural Rendering Techniques,
New York 1985.

Lin, Mike:
Drawing and Desiging with Confidence.
A step-by-step guide,
New York 1993.

Meuser, Natascha:
Construction and Design Manual.
Drawing for Architects,
Berlin 2015.

Porter, Tom / Goodmann, Sue:
Design Drawing Techniques,
Toronto 1991.

Porter, Tom:
How Architects Visualize,
London 1979.

Reid, Grant:
Landscape Graphics,
2002.

Sullivan, Chip:
Drawing the Landscape,
New York 1997.

Treib, Marc:
Representing Landscape,
New York 2008.

Treib, Marc:
Drawing/Thinking.
Confronting an electronic Age,
London 2008.

Walker, Theordore / Davis, David:
Plan Graphics,
New York 1990.

Wilson, Andrew:
The Book of Garden Plans,
London 2004.

Picture credits

Most of the sketches and examples shown in this book are by the author. All other picture credits are listed here. The author would like to thank all colleagues, students and alumni for their contribution to this book.
Should someone have been left out or a name be incorrect, the author kindly asks to be contacted in order to rectify any mistakes in upcoming editions.

14–15	Natascha Meuser
34	Table of projections by Prof. Justus Thyroff
38	Residential area/ site plan by Justus Thyroff / Wolf Rahm
41	Roof diagrams by Justus Thyroff
58	Thomas Grubert (bottom right)
59	Monika Supé
63	Martin Heiler (top right) / Annika Hepp (bottom left)
69	Matthias Schurer (left)
75	Thomas Grubert, Lisa Otten
78	Thomas Grubert
81	Thomas Grubert
83	Lisa Otten (curved pergola)
84	Pergola design by Stefan Schiessl (top left) / Ina Woeste (top right) / Andreas Kick (bottom left) / Pergola design by Julian Ulrich (bottom right)
88	Sabrina Jodoin (right)
89	Garden after Burle-Marx (left) / Angelika Reinhold (right)
94	Monika Supé (top right) / Maximilian Trautner (bottom right)
103	Merlin Bartholomäus
107	Britta Nickel (garden design, top right)
110	Boat house exercise by Wolf Rahm
120	Monika Supé (tree, top level, 2nd from left)
133	Toransicht nach Stefan Wenning
134	Student work (examples bottom left)
137	Alexandra Althaus (bottom)
144–145	Boat house exercise by Wolf Rahm
149	Sabrina Jodoin (middle) / Angelika Reinhold (bottom)
154	Monika Supé (top left)
155	Andreas Kick (middle)
169	Carolin Fischer (top left) / Lisa Otten (top right) / Sönke Küper (bottom left)
173	Julia Brüning
217	Chris Rascher (top left) / Selma Klophaus (middle right) / Andreas Steber (bottom left)
220	Monika Supé
225	Student work, Winter 2005 (top left)
226–227	Layout diagrams by Dr Monika Supé, Student work, Winter 2005 (227, bottom left)
228–231	Layout diagrams and examples by Dr Monika Supé
234	Monika Supé (top left)
238	Futura Alphabet by Wolf Rahm
239	Stefan Wenning (left) / Gunilla Lehmann (middle) / Student work, winter term 2007 (right)

The *Deutsche Nationalbibliothek* lists this publication in the *Deutsche Nationalbibliografie*; detailed bibliographic data are available online at http://dnb.d-nb.de.

ISBN 978-3-86922-852-5

© 2014 by DOM publishers, Berlin (1st Edition)
© 2016 by DOM publishers, Berlin (2nd Edition)
© 2021 by DOM publishers, Berlin (3rd Edition)
© 2025 by DOM publishers, Berlin (4th Edition)
www.dom-publishers.com

This work is subject to copyright. All rights are reserved, whether the whole part of the material is concerned, specifically the rights of translation, reprinting, recitation, broadcasting, reproduction on microfilms or in other ways, and storage or processing on databases. Sources and owners of rights are stated to the best of our knowledge; please point out any we might have omitted.

Proofreading
Laura Thépot
Mariangela Palazzi-Williams
Sabrina Wilk

Final Editing
Stefanie Villgratter

Design
Daniela Donadei

Cover Illustration
Masako Tomokiyo

Print
Tiger Printing (Hong Kong) Co., Ltd.
www.tigerprinting.hk